NAZI SCIENCE

MYTH, TRUTH, AND THE
GERMAN ATOMIC BOMB

NAZI SCIENCE

MYTH, TRUTH, AND THE
GERMAN ATOMIC BOMB

MARK WALKER

PLENUM PRESS • NEW YORK AND LONDON

Library of Congress Cataloging-in-Publication Data

On file

ISBN 0-306-44941-2

© 1995 Mark Walker
Plenum Press is a Division of Plenum Publishing Corporation
233 Spring Street, New York, N.Y. 10013-1578

10 9 8 7 6 5 4 3 2 1

Printed in the United States of America

Acknowledgments

I am very grateful to the many people who helped me with this project. Preliminary versions of Chapters 8, 9, and 10 appeared in *Vierteljahrshefte für Zeitgeschichte* (volumes 38 (1990), 45–74 and 41 (1993), 519–42), and Chapters 6 and 7, in *Historical Studies in the Physical and Biological Sciences* (volume 22 (1992), 339–89). I am indebted to John L. Heilbron and the editorial board of the *Vierteljahrshefte* for their support and editorial advice. Many of my colleagues read all or part of this book and gave me their helpful criticism: Philippe Burrinh, David DeVorkin, Michael Eckert, Paul Forman, Dieter Hoffmann, Ian Kershaw, Andreas Kleinert, Michael Neufeld, Ray Stokes, Carl Friedrich von Weizsäcker, Andreas Wirsching, and my Union College colleagues Steve Berk, Faye Dudden, Erik Hansen, and Teresa Meade. Finally, Linda Greenspan Regan, my editor at Plenum Press, greatly improved the manuscript through her insightful comments and suggestions.

This book could not have been written without the generous help of many archives and archivists: the Academy Archives of the Berlin-Brandenburg Academy of Sciences, the Berlin Document

Center, the Archives of the Humboldt University, the Archives of the Max Planck Society, and the State Prussian Library in Berlin; the State Archives in Hamburg; the Federal Archives in Koblenz and Potsdam; the German Museum, the Institute for Contemporary History, and the Max Planck Institute for Physics and Astrophysics in Munich; and the National Archives and Records Services in Washington, D.C. Special thanks go to Helmut Drubba for sending me a great deal of valuable information.

I received financial support for my research from the Alexander Humboldt Foundation, the Berlin Program of the Social Science Research Council, and Union College. Ulrich Albrecht, Andreas Heinemann-Grüder, and the Free University welcomed me as a guest in Berlin, as did Baudouin Jurdant, Josiane Olff-Nathan, and the rest of GERSULP at the University of Strasbourg. Perhaps most important is the support I have always received from the History Department at Union College.

My colleague Monika Renneberg and I have recently edited a collection of essays on science, technology, and National Socialism.[1] I cannot improve on the dedication we used in that book, so I would like to repeat it here.

> *This book is dedicated to all those critical voices who have tried to illuminate this ambivalent chapter of history, but were unappreciated, ignored, and discouraged.*

Contents

Contents

1

Introduction

Was there a peculiarly "Nazi science"? Were there uniquely "Nazi scientists"? These questions are deceptively simple. Even the term "Nazi" is frustratingly difficult to define. A minority of Germans, including National Socialist leaders like Adolf Hitler, Heinrich Himmler, and Josef Goebbels, were certainly "Nazis." An even smaller group, including the small circle of Army officers, churchmen, and aristocrats who tried and failed to assassinate Hitler in 1944, certainly resisted National Socialism. But the conduct and conviction of tens of millions of other Germans were not so clear cut. There is no simple definition for the term "Nazi." Mere membership in the National Socialist German Workers Party does not suffice: there are many examples of party members who opposed Hitler's murderous policies and of non-members who actively supported them. Instead, individual Germans have to be examined and judged on a case-by-case basis, and different observers may come to different conclusions.

It is important to distinguish between the minority of scientists who happened to be followers of Adolf Hitler and supporters

1

of National Socialism and the majority of scientists who placed their science in the service of the National Socialist state. The first group deserves criticism or even condemnation, but their support of National Socialism may have had little to do with their profession and therefore may tell us little about the peculiar and specific effect National Socialism had on science. The latter scientists may or may not have deliberately supported Hitler's political movement, but in some way their teaching or research was transformed, channeled, and exploited by the National Socialist state. An individual in this group is arguably far more interesting to the historian, for his story may help answer an important question in the history of science: can political ideology affect science? Was "Nazi science" different from contemporary Soviet or American science?

The fundamental problem for our understanding of science under National Socialism is the persistent and virulent use of the Janus-like combination of hagiography and demonization, the black-and-white characterization of scientists—like other professions and social groups—as fitting into three mutually exclusive categories: "Nazi"; "anti-Nazi"; or neither one nor the other. One could also label these categories "Heaven," "Hell," and "Purgatory," for they are based on the timeless, if sometimes simplistic theme of the struggle between good and evil.

A spectrum of "shades of gray" is far more useful than the black-and-white model for studying science and scientists under Hitler.[2] Although the two ends of this spectrum can also be thought of as "Nazi" and "anti-Nazi," these extremes are usually not reached, only approached. Almost every individual or institution in Germany embodied some elements that were either "Nazi," "anti-Nazi," or neither.

Thus for every scientist like the physicist Johannes Stark and the mathematician Theodor Vahlen, who clearly identified themselves with the National Socialist movement, there were far fewer scientists like Albert Einstein, who steadfastly opposed Hitler's movement (opposition facilitated because he left Germany before Hitler came to power). There were incomparably more scientists like Werner Heisenberg and Carl-Friedrich von Weizsäcker, who have been judged both "Nazi" and "anti-Nazi," and whose conduct during the Third Reich remains both controversial and open to interpretation.

This book will illustrate science during the Third Reich as a differentiated spectrum of shades of gray. Chapter 2 "The Rise and Fall of an 'Aryan' Physicist" and Chapter 3 "The Alienation of an Old Fighter" tell the story of the physicist Johannes Stark, an early supporter of Hitler and arguably one of the most prominent National Socialist scientists. Stark is perhaps best known for his infamous public attack on Werner Heisenberg calling him a "white Jew" in the SS newspaper *Das Schwarze Korps*. Chapter 2 begins in the last years of the German Empire and ends before the attack in *Das Schwarze Korps*, thereby explaining *why* Stark made his attack. Chapter 3 begins with Stark's concerted campaign of character assassination and ends with Stark's "denazification" after the war. These two chapters also investigate the history of the failure of the *Deutsche Physik* (literally translated as "German physics," sometimes translated as "Aryan physics") movement to win the support of the National Socialist state.

Chapter 4 "The Surrender of the Prussian Academy of Sciences" and Chapter 5 "A 'Nazi' in the Academy" tell the history of the transformation of this academy into a National Socialist tool. This "coordination" is often portrayed as an irresistible seizure of power by two National Socialist mathematicians, Ludwig Bieberbach and Theodor Vahlen. Chapter 4 begins in the Weimar Republic and ends with the academy's voluntary surrender to National Socialism *before* it elected Vahlen. Chapter 5 begins with Vahlen's entry to the academy and continues on into the postwar period. Together the two chapters focus on the persistent, courageous, yet ultimately futile efforts by the physicists Max Planck and Max von Laue to save the academy, as well as the gradual, step-by-step, and ultimately successful efforts of Bieberbach and Vahlen to undermine, control, and transform it.

Chapter 6 "Physics and Propaganda" and Chapter 7 "Goodwill Ambassadors" provide a thorough examination of the many foreign lectures the physicist Werner Heisenberg made during the

Third Reich, with the support of the National Socialist state and inevitably as a participant in its cultural propaganda. These lectures include the controversial visit Heisenberg paid to his Danish colleague Niels Bohr in occupied Denmark in September of 1941. Chapter 6 covers the period from the start of the Third Reich to the height of German military success in late 1941, ending with the Bohr visit. Chapter 7 begins with the winter of 1941–1942, when the war began to go sour for Germany, and finishes with the end of the Third Reich. These two chapters illustrate how ambivalent and ambiguous it was for scientists to work within the National Socialist system, regardless of what they did or what their intentions were.

"Hitler's Bomb" surveys the history of the wartime uranium research project, including the background of science during the Weimar Republic and under National Socialism. "The Crucible of Farm Hall" examines the postwar internment of ten German nuclear scientists in an English country house, Farm Hall, where their captors secretly listened to their conversations and where they first learned of the bombing of Hiroshima. The pressure of events and enforced isolation made Farm Hall into a crucible, where the first myths surrounding the German atom bomb were forged. "The Myth of Hitler's Bomb" examines these persisting postwar myths and legends, which have changed over time.

Finally, this book closes with an investigation of a taboo of modern science: the scientist as "Fellow Traveler." If we want to understand how National Socialism affected German science, we cannot restrict ourselves to the few scientists who enthusiastically embraced the Third Reich, and those even fewer scientists who actively and consistently resisted it. Instead we must also include those very many scientists who neither resisted nor joined Hitler's movement, rather who went along for the ride.

2

The Rise and Fall of an "Aryan" Physicist

Without a doubt Johannes Stark is one of the most famous and infamous "Nazi" scientists. His Nobel Prize, irascible nature, and often vicious ideological attacks on modern physics and physicists make him both an intriguing subject and the perfect villain. Stark is perhaps best known for his infamous attacks on Werner Heisenberg, labeling him a "white Jew" in the *Schutzstaffein* (SS) newspaper. But there is much more to this story. Therefore Stark's relationship with National Socialism will be broken up into two chapters, "The Rise and Fall of an 'Aryan' Physicist," which ends before the attack on Heisenberg, and "The Alienation of an Old Fighter," in order to place his attacks on Heisenberg into context. Stark's successes, but especially his failures, during the Third Reich tell us a great deal about the interaction of physics and National Socialism.

☩   ☩   ☩

The Weimar Republic Stark was a talented and ambitious physicist. In 1909 he took up his first professorship at Aachen. The outbreak of World War I transformed him spiritually and ideologically into an extreme German nationalist.[3] Although Stark may have been more extreme than most of his colleagues, in general, German scientists did rally uncritically behind the German war effort. Professional setbacks also influenced his development. Stark's relationship with the Munich theoretical physicist, Arnold Sommerfeld, degenerated into a bitter and unprofessional polemic over physics, which formed the basis for their subsequent antagonism. When Stark's hopes of being called to a professorship in Göttingen were dashed by the appointment of Sommerfeld's student, Peter Debye, in 1915, Stark blamed the "Jewish and pro-Semitic circle" of mathematicians and theoretical physicists there and its "enterprising business manager" Sommerfeld.[4]

In 1917 Stark moved on to Greifswald, where he experienced the revolution that followed the German defeat. The German surrender in the fall of 1918 took most Germans by surprise since their government had fed them propaganda, promising that victory was at hand. When the soldiers returned home—often with their weapons—they found a home front devastated by hunger and a power vacuum. Throughout Germany left-wing soldiers' and workers' councils took over political power at the local level. Many Germans believed that the country was going to experience a repeat of the Russian Revolution. Right-wing militias were formed to avert the Communist threat, plunging the country into a short, bloody civil war. An unlikely alliance between the German military and the Social Democratic Party with a new constitution in 1919 eventually brought some political stability, but not until many had died and a great deal of resentment had been caused.

The atmosphere of Greifswald, a small university town, and in particular the extremely conservative and nationalistic faculty and student body appealed to Stark. When the socialists gained power in Greifswald, Stark actively opposed them and thereby began his political career as a German nationalist and conservative long before anyone had heard of Adolf Hitler or the National Socialist German Workers Party (NSDAP).[5]

In 1920 Stark received the 1919 Nobel Prize for his discovery of the Stark effect—the splitting of spectral lines in an electric field—and moved on to the University Würzburg in his native Bavaria. He now became more active in the politics of the physics community. Berlin physicists, who tended to be more liberal, cosmopolitan, and theoretical, dominated the German Physical Society and had alienated more conservative physicists from other parts of Germany. In April 1920 Stark began soliciting members for his alternative German Professional Community of University Physicists, an organization Stark intended to dominate physics and control the distribution of research funds.

But Stark's efforts were thwarted. The Physical Society mollified most conservative scientists by electing as president, Wilhelm Wien, one of their number who was much easier to deal with than Stark. The two main funding organizations, the private Helmholtz Foundation and the state-run Emergency Foundation for German Science (*Notgemeinschaft der Deutschen Wissenschaft*, henceforth NG), also preserved their independence by lining up influential scientists and patrons.[6] When Stark realized that his voice would be only one among many setting science policy, he withdrew. Stark's efforts in 1920 were a preview of the action he would take with political backing at the beginning of the Third Reich.

Scientific opposition to portions of modern physics, and in particular to theoretical physics, took on a more ominous tone in the early twenties. In 1921 Wilhelm Wien recognized that the general theory of relativity was engulfed by an unprecedentedly bitter and sometimes unprofessional debate which had left the realm of science and become entangled with politics and dogma.[7] Indeed it was considered good form in the twenties for a scientist to distance himself from the political and ideological battles if he wanted to comment critically on Albert Einstein's work.[8] Ironically, the postwar anti-Semitic attacks against Einstein as creator

Albert Einstein, 1922. (From the Burndy Library, Courtesy of the AIP Emilio Segrè Visual Archives.)

of the theory of relativity were an inversion of wartime foreign chauvinism. Einstein's work, the type of science which the French had criticized as typically "German" physics during World War I, was criticized by right-wing German conservatives as typically "Jewish" after 1919.[9]

Philipp Lenard, fellow Nobel laureate and professor of physics at the University of Heidelberg, was the first prominent German scientist to attack "Jewish physics" and call for a more

Philipp Lenard, 1936. (From Ullstein Bilderdienst, Courtesy of the AIP Emilio Segrè Visual Archives.)

"Aryan" physics. In 1922 he published a word of warning to German scientists, accusing them of betraying their "racial allegiance" and noting that the transformation of an objective question into a personal fight was a "known Jewish characteristic."[10]

Lenard's arguments against the theory of relativity initially had nothing to do with anti-Semitism or personal antagonisms. Indeed Lenard had followed Einstein's career from the beginning with benevolent interest, calling him a deep, comprehensive thinker in 1909. Lenard's opposition to the theory of relativity began in 1910, but did not include personal attacks on Einstein. As late as 1913 Lenard was toying with the idea of calling Einstein to a professorship of theoretical physics in Heidelberg. The discussion between the two physicists became sharper during the war, but remained within the bounds of scientific debate.[11]

In 1920 a popular lecture series sponsored by the "Working Group of German Scientists for the Preservation of Pure Science" opposed to Einstein's theory of relativity was held in the Berlin Philharmonie. This organization probably never existed, except on paper, and was the invention of the fanatical Einstein opponent Paul Weyland.[12] Einstein's subsequent reply, "My Answer to Anti-Relativity Co.," appeared in a Berlin daily newspaper. His unfinished question, "if I would be a German nationalist with or without swastika instead of a liberal, internationalist Jew...," cut to the heart of the matter and raised the stakes in the debate.[13]

Before 1920 most physicists had taken care to keep their criticism of Einstein well within the bounds of professional discourse.[14] Einstein's supporters were the first respectable scientists explicitly to use the word *anti-Semitism*, and ironically gave their opponents the opportunity to claim that it was Einstein who had introduced race and religion into a scientific debate. However, the floodgates were now opened.

Lenard began to incorporate anti-Semitism into his publications against Einstein and his theory in 1921. The lost war was certainly part of the reason, but perhaps just as important was Einstein's public criticism of Lenard in the aftermath of the anti-Einstein conference. Although Lenard had not taken part in the Berlin lectures and hitherto had only expressed his opinion in a professional fashion in scientific journals, Einstein's personal attack in the daily press deeply offended Lenard, who was seventeen years his senior.[15] When Lenard refused to lower his institute flag after the assassination of Walther Rathenau, a Jewish German foreign minister and friend of Einstein, the conservative physicist was attacked and publicly humiliated by a mob. This experience was an important factor in Lenard's turn towards more blatant racism and anti-Semitism.[16]

Ludwig Glaser, one of Stark's advanced students, was an ambitious and competent scientific entrepreneur, who edited his own technical journal and ran his own laboratory, which specialized in physical and chemical special investigations (optics, metallurgy, spectral analysis) as well as the assessment of patent applications and used scientific equipment. More importantly, Glaser was a convinced and determined opponent of Albert Einstein's theory of relativity. He had taken part in the Berlin conference, and thereby became personally involved in the controversy surrounding Einstein.

According to Max von Laue, an expert on the theory of relativity and friend of Einstein, Glaser restricted himself to professional arguments in his Berlin lecture, even though he did not succeed in convincing Einstein's scientific supporters. Von Laue only faulted him for being too one-sided.[17] In contrast, Glaser complained about the demagogic, personal, and unscientific attacks made against the Berlin lecturers at the subsequent convention of German scientists in Nauheim.[18] Glaser published several articles against Einstein's theory and called the expectations held by supporters of the theory of relativity premature and exaggerated. Stark's student stuck to scientific arguments, just like Lenard had at first. During the Weimar Republic there was no trace of the virulent anti-Semitism Glaser developed during the Third Reich.[19]

In the summer of 1921 Stark accepted a *Habilitationsschrift* (a sort of second Ph. D. thesis) from Glaser on the optical properties of porcelain. His Würzburg colleagues questioned whether such a topic really constituted a scientific advance. Some mocked Glaser's thesis as a "doctor of porcelain." However, objections were also raised because of Glaser's ties to the anti-Einstein group, and his participation in the Berlin conference had aroused such deep bitterness.

Stark considered the academic opposition to Glaser part of a conspiracy by Einstein's supporters. Furious, Stark resigned, returned to his original home, and invested his Nobel Prize money in various industrial enterprises. Almost immediately, Stark regretted his decision to resign. He probably expected to be given the presidency of the Imperial Physical-Technical Institute (PTR), the German equivalent to the National Bureau of Standards, a promotion which would have allowed him to stay in the academic physics community. When he was passed over and thereby isolated, his bitterness grew.[20]

If Einstein's scientific theory and support for internationalism, pacifism, and the Weimar Republic had not made him controversial enough, then the Nobel Prize he received in 1922 made him a target for vindictive abuse and attack from the radical right. Stark was now alienated if not enraged by Einstein's political stance.[21] Stark's 1922 book, *The Contemporary Crisis in German Physics*, attacked modern physics—roughly speaking, quantum mechanics and relativity—as "dogmatic."

Although this argument did not yet include anti-Semitism, Stark did criticize *how* the theory of relativity was being propagated by Einstein and others. According to Stark, Einstein and his supporters had improperly publicized his scientific theory through newspaper articles and foreign lectures. Since the propaganda for Einstein's theory spoke of a revolution in science, Stark noted, it found fertile ground in the postwar period of political and social revolution. Einstein had betrayed Germany and German science with his internationalism.[22]

Stark's book did not go over well. Max von Laue's pointed review, which publicized the personal antagonism which now existed between Stark and himself, drew the battle lines for the subsequent struggle over Einstein's science: on one side, scientific support of the theory of relativity and opposition to the racist, political, and ideological attacks against its creator; on the other side, escalating personal attacks on Einstein and his work which had less and less to do with science and more and more to do with the National Socialist movement.[23]

The long-standing cordial personal and professional relationship which Lenard and Stark had enjoyed now became a political collaboration. Both scientists began to engage in political activity only after their professional work had diverged from the main path taken by modern physics.[24] Although they opposed all or part of quantum mechanics and the theory of relativity, for Lenard the distinction between "Aryan" and "Jewish" science was a matter of ideology; for Stark it was a weapon to use against those who had kept him a pariah for so long.[25]

The *Deutsche Physik* movement they founded was the result of three different factors: the opposition of professionally conservative scientists to modern physics, often because they were not in a position to understand, appreciate, or use it; the opposition of anti-Semitic scientists to Einstein, other Jewish scientists, and the physics they created; and the opposition of right-wing, nationalistic scientists to the pacifist, internationalist stand taken by Einstein as well as his support of the Weimar Republic. When the three groups of professionally conservative, anti-Semitic, and nationalistic scientists overlapped, they formed *Deutsche Physik,* a political movement composed of scientists using the rhetoric of science. These physicists had nothing new to offer in the way of science, and are best characterized by what they rejected: modern theoretical physics, especially quantum mechanics and the theory of relativity, all of which came increasingly under the heading "Jewish physics."

The anti-Semitism of *Deutsche Physik* fit well into the political climate of Weimar Germany. As early as the autumn of 1923, in the aftermath of the "Beer Hall Putsch," Stark had publicly supported Adolf Hitler and his National Socialist movement. In November of that year Hitler had led a march from a beer hall in Munich designed to topple the city government in a coup and eventually lead to a national revolution modeled on Mussolini's successful march on Rome. The coup ended when Bavarian police fired on the marchers. Although Hitler did not distinguish himself by bravery when the march collapsed, he regained his composure at his trial for treason. Hitler managed to turn his trial into political propaganda, admitting guilt but rejecting the idea that his attempt to topple the Weimar Republic was a crime. His right-wing judges were sympathetic and gave him the most lenient sentence possible—five years with the understanding of early probation.[26]

Johannes Stark, 1931, from his NSDAP party book. (From the Berlin Document Center.)

A year later, while Hitler was serving time in Landsberg prison for his part in the failed putsch, Stark and his wife invited him to recuperate with them after his release, an offer for which Hitler thanked him heartily.[27] In May 1924, Lenard and Stark published an open letter supporting Hitler. Their mystical prose fit well into the Aryan-supremacist rhetoric of the day:

> ...the struggle of the spirits of darkness against the bearers of light ... [Hitler] and his comrades in struggle ... appear to us as gifts of God from a long darkened earlier time when races were still purer, persons still greater, spirits still less fraudulent.[28]

Lenard's and Stark's overt support for National Socialism was unusual among academicians and rare among physicists.[29] Hitler was very grateful for the public support of two leading German scientists, coming as it did at a precarious time for his movement.

Stark joined the NSDAP (National Socialist German Workers Party) in 1930. He earned the title of an "Old Fighter" for Hitler's movement—someone who had joined before Hitler was appointed Chancellor in 1933, i.e., someone who could not have been a political opportunist.[30] National Socialist ideology was congenial to Stark, but his early activism for the National Socialists has an additional explanation: Stark found in National Socialist circles the honor and recognition as an important scientist that his fellow academics had denied him, despite his Nobel Prize.[31]

Stark was even willing to stop his scientific work in order to help Hitler in the National Socialist leader's final struggle to gain power. After Hitler emerged from prison and refounded the NSDAP, he proclaimed that he would henceforth take the "path of legality." In practice, this meant that the National Socialists would not try to seize power in Germany via a coup, but instead would work within the constitution as a political party. Hitler and other leading National Socialists often stated openly that, although they intended to come to power legally, once in power, they would tear up the constitution and end democracy. At the time few people took this threat seriously.

During the last three years of the Weimar Republic (1930–1932), the NSDAP mounted what amounted to a perpetual election campaign. In 1932 there were three national elections: two for parliament and one for the presidency. The National Socialists were successful in large part because of the many dedicated members of their movement like Stark who mobilized voters at the local level, by writing political pamphlets and organizing and leading mass rallies. In 1932 Stark agitated for the National Socialist movement near Traunstein and his estate and repeatedly held large public meetings in the area.[32] Hitler himself thanked Stark for his efforts on behalf of the NSDAP.[33] By the end of the Weimar Republic, Stark, who owned an estate in rural Bavaria, was seen by the population as a spokesman for the National Socialist party.[34]

But from the very beginning, Stark was fundamentally ambivalent about the radical right. In the early twenties Stark told Lenard of his pessimism in regard to politicians on the far right. They were profiteers, ambitious, and rowdies. Although the National Socialist movement was his last hope for the resurrection of the German people, his optimism was vanishing and being replaced with a profound pessimism.[35] Stark seems to have shared a common attitude among supporters of Hitler's movement: he was disturbed by the behavior of the so-called "little Hitlers," the low-level National Socialist officials, but nevertheless simultaneously embraced Hitler himself as leader of the movement with uncritical admiration and trust. Hitler was aware of the credibility gap between himself and his party and both cultivated and exploited it: whenever there was credit to be taken, he took it; whenever things went wrong, the blame fell on the little Hitlers in the party.[36]

✠   ✠   ✠

The Third Reich The subsequent step-by-step "coordination" of every aspect of German society which followed Hitler's appointment as German Chancellor was unsettling if not deeply disturbing for most German physicists.[37] More than 15 percent of all academic physicists emigrated willingly or unwillingly after 1933, although the actual damage to physical research was much greater than this number implies.[38] Prestigious scientific research institutions like the semi-private Kaiser Wilhelm Society (KWG) (established early in the twentieth century in order to facilitate

Johannes Stark, 1933. (Courtesy of the Ullstein Bilderdienst.)

research outside of the universities) "coordinated" themselves in the hope of avoiding even tighter control from the National Socialist government.[39]

The National Socialist revolution effectively purged the civil service of potential opponents to the new regime. Since all university employees were civil servants, this policy also purged German physics of "non-Aryans" and leftist scientists.[40] But that was not enough for the small group of physicists gathered around Lenard and Stark. They wanted to control all future university appointments, scientific publication, and funding of research. In other words, they wanted a "second revolution" in German physics in order to accomplish what Lenard and especially Stark had failed to achieve in Weimar.[41]

Within a week of Hitler's appointment to German Chancellor, Stark enthusiastically wrote Lenard that the time had finally come when they could implement their conception of science and research. Stark used the opportunity of a congratulatory letter to his personal acquaintance, National Socialist Minister of the Interior Wilhelm Frick, to tell him that Stark and Lenard would be pleased to advise him. Stark had specific help in mind. He wanted the prize that had eluded him in the early twenties—the presidency of the Imperial Physical-Technical Institute.[42]

Lenard went directly to Hitler and offered his services. There was a great deal to be done, Lenard told him, for the entire university system was in badly rotted condition. Although there were not enough really talented scientists to fill the openings, Lenard could find enough thoroughly German physicists who were good enough. Lenard himself was ready to help in checking, evaluating, influencing, and if necessary, rejecting and replacing candidates.[43]

At first it appeared that the two leaders of *Deutsche Physik* would get their wish. In July 1933 von Laue complained to his colleague Walther Gerlach that his influence was now insignificant. To get something done one had to go through Lenard and Stark.[44] By November Lenard and Stark had been promised that they would be consulted before scientific professorships were filled.[45] Stark's almost boundless ambitions extended to the KWG, where he hoped that Max Planck, the current KWG president, would be forced to resign and make way for a National Socialist. But Stark first asked Lenard if he wanted the job.[46] His colleague replied that he was only interested in squashing and then completely rebuilding the society.[47] Stark was sympathetic. He did not want to take over the KWG presidency himself, but was very interested in the Emergency Foundation and distributing its considerable funds for scientific research.[48]

Stark made his intentions for German science public at the September 1933 meeting of the German Physical Society in Würzburg. According to von Laue, Stark practically declared himself the dictator of physics. Many of his listeners found most disturbing Stark's plans for the physics publishing business. He wanted a general editorship for all physics journals, which would decide whether or not work would be published and in which journal it would appear. This editorship would, of course, be under his personal control. In effect, Stark was merely advocating the type of totalitarian control that Josef Goebbels' Reich Cultural Chamber had over newspapers and general literature, and which had become common in the Third Reich.

Von Laue and others rightly feared that if Stark's plan succeeded, then certain types of theoretical physics would effectively be silenced in Germany. The Würzburg conference probably reminded Stark of his self-inflicted professional isolation during Weimar, and he did not mince words: if the publishers did not go along, then he would use force. Although his plans certainly appeared to be a threat to intellectual and scientific freedom, Stark went out of his way to deny this in his Würzburg speech, either because he was employing the common but often effective National Socialist tactic of falsehood, or because in his own mind, "freedom of research" meant scientists were free only to do the sort of research he valued.[49]

If Stark had hoped for the quiet acquiescence of his scientific colleagues, he was disappointed. Von Laue challenged him publicly at Würzburg by an implicit yet clear comparison of the contemporary fight against the theory of relativity with the Catholic church's trial of Galileo and subsequent attempts to ban the Copernican model of the planets moving around the sun. When von Laue noted that the earth still moves, his listeners knew exactly what he meant: despite the rhetoric of *Deutsche Physik*, the theory of relativity was true.[50] Stark was enraged by von Laue's speech, and subsequently reported to National Socialist officials that von Laue had received the enthusiastic applause of all the "Jews and their fellow travelers present."[51] For his part, von Laue had carefully not attacked the National Socialist government or even National Socialism, rather the *Deutsche Physik* campaign against Einstein.

The first tangible fruits of Stark's long-standing support for Hitler's movement came in May 1933, when he was appointed president of the Imperial Physical-Technical Institute—despite being rejected unanimously by the scientists consulted.[52] Stark had been waiting for more than a decade for this opportunity. He threw himself into plans for an extensive reorganization and massive expansion of the PTR and took steps to ensure a more National Socialist institution. However, the PTR administration had already fired all its Jewish employees in April, before Stark became president.

Stark did cut off certain lines of basic research associated with modern physics, although much valuable research continued. The Institute took on a distinctly National Socialist flavor when Stark implemented the "leadership principle." Each individual had a specific position in a strict hierarchy. He had to follow all orders received from above without question, but in turn could expect unquestioning obedience from anyone below him. In the summer of 1933 the new PTR president fired the "Jews and leading figures of the previous regime" from the PTR advisory committee, which itself soon disappeared as well.[53]

The new president had gigantic, if not absurd, plans for an expanded PTR, including fifty large institutes, three hundred labs, and thousands of scientific workers. Initially Stark was able to win Hitler's personal support for his plans. However, the proposed move to Munich or Potsdam fell victim to bureaucratic in-fighting, the passive opposition of the Reich Ministry of Finance, and shortage of funds. Nevertheless Stark did expand the PTR significantly, concentrating on military or military-relevant research.

In his infamous speech in Würzburg, Stark trumpeted that the new PTR would have great importance for science, the economy, and the national defense. A memo he wrote at the same time described the PTR as a central organ providing scientific support for the entire economy and national defense. By 1937 the PTR was working closely with the military, especially the Air Force and

Army Ordnance. The PTR had originally been created to establish national standards for science and technology; it now set the standards for armaments of all types, thereby taking on a key responsibility for the armed forces. Such a concentration on military research inevitably meant that there was less time and resources for basic research.[54]

There was not enough money to go around in the Third Reich. At first science was not a high priority for the National Socialists, so Stark almost immediately encountered personal and bureaucratic resistance to his ambitions. In October 1933, Stark asked the NG for 200,000 Reich Marks (the official exchange rate paid 4.2 "Gold Marks" to the dollar) in order to begin accelerated research important for the economy and rearmament. Moreover, he argued that physical research throughout Germany had to be organized and channeled into the national defense.[55]

An inter-ministry meeting was called to discuss Stark's exceptional request and included Erich Schumann from the Defense Ministry, representatives from the Finance and Interior Ministries, and Friedrich Schmidt-Ott from the Emergency Foundation. The official from the Ministry of the Interior began by asking Stark for precise details of the tasks to be funded. Stark responded instead with a long presentation in which he argued that a series of investigations had to be started immediately in the interest of national defense. He needed several hundred thousand Reich Marks, although at the moment Stark admitted that he could not provide a precise budget.

Schumann responded that most of this work was already being carried out elsewhere under the authority of the Army. Schmidt-Ott added that other projects mentioned by Stark were being done by the Transport Ministry with funding from the Emergency Foundation. Indeed all the subjects mentioned by Stark were already being examined, either by the Army Ministry, the Transport Ministry, Ministry for Aviation, the Postal Ministry, or the national Train Company. Stark responded by promising to submit a detailed written proposal.[56]

The Ministry of Interior decided that this request could not be granted for legal reasons alone, never mind the fact that Stark's similar request in July for 100,000 Reich Marks had already been refused. The president of the PTR was clearly planning to use his institute to streamline and centralize research in Germany as much as possible, even though the other bureaucrats saw no need for a third such institution alongside the Emergency Foundation and KWG. The ministerial officials concluded from this case and others that Stark wanted to extend the influence of his institute further than was necessary. If Stark wanted funds, they decided, then he should apply to the Emergency Foundation like everyone else.[57]

Such internal bureaucratic conflict was typical of the "polycratic" nature of the National Socialist state. Despite the National Socialist rhetoric of a disciplined government organized along the lines of the leader principle, Germany in fact now consisted of several power blocs which both cooperated and competed for power.[58] Apparently Stark never bothered to submit the promised description of his proposed research program.[59] Even though ultimately Stark somehow managed to go over the heads of these bureaucrats and receive the money he wanted, this episode made clear how and why he was making many enemies among the National Socialists now running the state bureaucracy.[60]

Moreover, Stark's ideological enemies and half-hearted party comrades sometimes worked together against him. When the Prussian Academy of Sciences (PAW) considered admitting Stark in the late autumn of 1933, his old adversary, von Laue, managed to abort the nomination. Some governmental officials did push Stark's candidacy, but others in the Reich Ministry of Education (REM), who could have forced the PAW to admit the physicist, chose not to interfere.

Stark found time to continue his fight against modern physics, but at first he focused more on international opinion. In late 1933 Stark advised REM that a new debate over Einstein's theory of relativity in Germany would be superfluous, claiming that the scientific community had already made up its mind and there was hardly any more interest in such a debate.[61] Shortly thereafter,

Stark took his case against "Jewish" science to the readership of the prestigious British scientific weekly *Nature*. Stark's letter to the editor asserted that the National Socialist government had not directed any measure against the freedom of scientific teaching and research. On the contrary, Germany's new leaders wanted to restore this freedom, which had been restricted by the preceding democratic government. The political measures which had been taken against Jewish scientists and scholars were necessary, he argued, in order to curtail the great influence they had but did not deserve.[62]

The subsequent critical letters to the editor provoked another letter from Stark, a curious mixture of falsehood and insight. The National Socialist government had not persecuted Jewish scientists or forced them to emigrate, he insisted. It had merely reformed the civil service, including all kinds of officials, not just scientists. No government, Stark asserted, could be denied the right to reform its own civil service, and no group of officials, including scientists, could be granted an exception to such a law.[63] Stark was dishonest about the treatment of Jewish scientists, but he was right to point out that what was happening in Germany was not directed against science in particular. The "non-Aryan" scientists who lost their jobs and often were hounded out of Germany were persecuted because they were Jewish or for political reasons, not because they were scientists.

Stark took care to report his international propaganda efforts to the responsible German officials, noting that the National Socialist campaign against Jewish influence in German culture had provoked a strong response by Jews all over the world. Moreover, Stark added, the friends of Jewish scientists were trying to influence influential figures in the German government by arguing that Jewish scientists and especially their "Aryan" friends and allies in Germany had to be treated gingerly in order to pander to foreign opinion. Stark was mainly interested in using this opportunity to attack his favorite enemies, including the "sponsors of scientific Jewry" and friends and sponsors of Einstein who remained in their influential positions, specifically KWG president Planck, Berlin university professor von Laue, and Munich university professor Sommerfeld.[64]

But Stark did not attack all of his "non-Aryan" colleagues. In late 1934 the National Socialist Teachers League contacted Stark with regard to the experimental physicist Gustav Hertz. They wanted a scientific, pedagogic, political, and character assessment, and were especially interested in information regarding his *momentary* indispensability.[65] Stark responded that there was nothing Jewish about Hertz's statements, conduct, or scientific activity. In Stark's opinion, he was one of the few first-class German physicists, a Nobel laureate, and the nephew of the great physicist Heinrich Hertz. It would be stupid, Stark argued, to remove Hertz's right to teach just because his grandfather was a Jew. Moreover, Stark was convinced that Hertz would not take such humiliation quietly, rather would go abroad where he would be welcomed with open arms.[66]

Hertz lost his professorship nevertheless, and retreated into a research position in German industry, where during the war he devoted himself to military research. Stark subsequently went out of his way to assist Hertz and his co-workers.[67] Stark was certainly anti-Semitic, but the Hertz affair illustrates that there is more to the story. Like many people during the Third Reich, Stark made his own definition of who was or was not a "Jew." Thus Stark could both assert that someone like Hertz was not really "Jewish" even though he fell under the legal definition of "non-Aryan" used by the National Socialists (having a grandparent who had belonged to the Jewish religious community), and attack others who were legally "Aryan" as "Jewish in spirit." However, the fact that Stark's racism was sometimes opportunistic does not make it any better. His anti-Semitism nevertheless remained virulent and vicious.

Stark did not always take the initiative himself in his efforts on behalf of National Socialism. In the summer of 1934 a high-ranking official in the Ministry of Propaganda suggested that Stark arrange a public declaration of support for Adolf Hitler by the twelve "Aryan" German Nobel laureates.[68] Stark sent telegrams to his fellow laureates and asked them to sign the following text: "In

Adolf Hitler we German natural researchers perceive and admire the savior and leader of the German people. Under his protection and encouragement, our scientific work will serve the German people and increase German esteem in the world."[69]

Werner Heisenberg's return telegram tried to refuse without saying no. Although he personally agreed with the text, he considered it improper for scientists to make political statements and therefore he refused to sign.[70] The rest of the laureates responded similarly. Stark reported his failure to Goebbels himself and went out of his way to damn his colleagues while underscoring his own zeal by forwarding on his colleagues' answers as well as his criticism of their unwillingness to help the National Socialist cause.[71]

Stark's greatest assets were his few direct lines of communication to the highest levels of the National Socialist state. On 30 April 1934, Stark sent an outline of his proposals for the reorganization of German science directly to Hitler. The Reich Research Council he proposed would set guidelines for all research, control all funding, and oversee all research institutions.[72] Less than a year later, the head of the Reich Chancellery, Hans Lammers, invited Stark to assess the organization of German research.[73] Shortly thereafter Stark tried to enlist the support of the Army for his plans to give the PTR a monopoly over technical testing and standards.[74]

Initially, Stark's lobbying paid off. In the spring of 1934 he was appointed the president of the German Research Foundation (DFG), the renamed successor to the Emergency Foundation and the clearinghouse for most governmental funding of scientific research. When Minister of Education Bernhard Rust fired the foundation president Schmidt-Ott, he told him that Hitler had personally ordered Stark's appointment.[75] Stark happily told Lenard that together they could now develop the universities and scientific research in a Germanic sense.[76] Indeed this appointment had an immediate effect on physics: Stark stopped funding theoretical work after he became head of the Research Foundation, and henceforth only funded certain types of experimental research.[77]

Lenard congratulated his colleague and celebrated the success of *Deutsche Physik* in the pages of the National Socialist daily *Völkischer Beobachter* (literally translated as "The People's Observer"):

It had grown dark in physics ... Einstein has provided the most outstanding example of the damaging influence on natural science from the Jewish side ... One cannot even spare splendid researchers with solid accomplishments the reproach that they have allowed the 'relativity Jews' to gain a foothold in Germany .... [The] theoreticians active in leading positions should have watched over this development more carefully. Now Hitler is watching over it. The ghost has collapsed; the foreign element is already voluntarily leaving the universities, yes even the country.[78]

Lenard's article is typical of the tactics employed by *Deutsche Physik* in that he simply asserted without any proof that the "relativity Jews" had threatened German science and Germany itself.

Unfortunately for Stark, his two presidencies were offset by other developments in National Socialist science policy. Stark had enjoyed excellent connections to Interior Minister Frick, but in August 1934 responsibility for scientific research was transferred from his ministry to Bernhard Rust's REM.[79] Henceforth, Stark would see many of his efforts to reorganize and control German science sabotaged, diverted, or taken over by hostile REM bureaucrats.[80]

Early in 1935 Stark was forewarned of an intrigue against him by an unexpected source. On 26 January KWG president and—using Stark's own label—"friend and sponsor of Einstein" Max Planck was called in by Rust, who read Planck part of an anonymous letter accusing Stark of making derogatory remarks to "non-Aryan" scientists about the policy of the Reich government. Such "anonymous" letters were often fabricated by the National Socialists themselves. Rust then asked Planck if he knew of such remarks and whether Stark had discussed the matter with him. Planck replied with great care that he would have to describe the account given in the letter as tendentious.

The Education Minister then directed Planck to put down in writing the facts as he knew them. According to Planck, Stark had remarked that, with regard to the effects of the "Aryan paragraph" in the new civil service law, which effectively fired all Jewish civil servants, in a few cases a somewhat milder process would be desirable in the interest of science. Moreover, Planck told Rust that he agreed with this opinion. Within a few days of his audience with Rust, Planck brought this matter to Stark's attention. If someone tried to use Planck's letter against Stark, then he now would know precisely how and why.[81]

This episode is significant for three reasons: it illustrates bureaucratic intrigue in the Third Reich; it demonstrates how scientists like Planck were exploited in such intrigues; and it, along with the Gustav Hertz affair, reveals that despite his *Deutsche Physik* rhetoric, Stark was willing to make exceptions when it came to his "non-Aryan" colleagues. Yet the few examples of Stark's compassion are outweighed by the much more common and prominent vindictiveness he showed to his self-appointed enemies.

The most prominent scientist attacked by Stark as "Jewish in spirit" was the young theoretical physicist Werner Heisenberg, the student of Stark's hated rival Sommerfeld and one of the creators of the quantum mechanics, in other words, of part of "Jewish physics." At first Stark did not single out Heisenberg for abuse like von Laue, Sommerfeld, or Planck. Since the latter three physicists had influential positions in German science, they stood in Stark's way; Heisenberg did not. That all threatened to change dramatically when Sommerfeld announced his retirement and the University of Munich requested Heisenberg as his replacement. The "Sommerfeld succession"[82] quickly was politicized and made into a prestige object in the struggle between "Jewish" and "Aryan" physics.

In the summer of 1934, when it appeared that Sommerfeld's Munich chair in theoretical physics would soon become free, National Socialist officials connected with the University of Munich contacted Stark and asked for his assistance in finding a suitable successor. Since the university faculty was under the influence of "pro-Semitic" forces, the party officials would be grateful if Stark could name a productive and militant National Socialist.[83] Stark responded immediately that the Munich appointment was very important to him.[84]

But Lenard's and Stark's desire to control university appointments and fill them only with candidates they found acceptable was complicated by their almost universal contempt for German physicists. In 1934 Lenard could hardly name ten physicists who would be suitable for science in the Third Reich.[85] Stark agreed wholeheartedly and argued that a professorship should be left vacant rather than be filled with the wrong person.[86] Finally, in a taste of what was to come, when Stark first tried to influence the Munich appointment in 1934, his party comrade and REM bureaucrat Theodor Vahlen politely declined, cynically arguing that regulations forbade any outside intervention in the search to fill a professorship. What Vahlen really meant was that only REM personnel would be allowed to manipulate and influence such matters.[87]

Lenard and Stark now began spreading their gospel in other ways. Lenard's four volume textbook on *Deutsche Physik* (1935)[88] argued that everything created by man, including science, depends on blood and race. Thus the Jews had developed their own physics, which was very different from *Deutsche Physik*—which, Lenard noted, could also be called "Aryan" or "Nordic" physics.[89] Jewish physics could best be characterized by the work of its most outstanding representative, the "pure-blooded Jew Albert Einstein" and his theory of relativity.[90]

The pompous renaming of the Heidelberg physics institute as the Philipp Lenard Institute in December 1935 provided an opportunity for Stark to rail against Jewish physics and Heisenberg.[91] Einstein had now disappeared from Germany. But unfortunately his German friends and supporters were still active in his spirit: Einstein's main supporter Planck was still president of the

KWG, and his interpreter and friend Max von Laue was still the physics expert in the Prussian Academy. And Heisenberg, "spirit of Einstein's spirit," Stark noted pointedly, was supposed to be distinguished by an academic appointment.[92]

Part of Stark's speech was subsequently used by a physics student named Willi Menzel in an article in the National Socialist newspaper *Völkischer Beobachter*: Einstein's theory of relativity, Heisenberg's matrix mechanics, and Schrödinger's wave mechanics were all dismissed as opaque and formalistic.[93] Heisenberg recognized the seriousness of Menzel's article and wrote his own piece for the National Socialist daily. But his article was accompanied by a counterattack by Stark. Heisenberg was still advocating "Jewish physics," and indeed expected that young Germans should take Einstein and his comrades as role models.[94] From this point onward, Heisenberg was the focal point for Stark's attacks on "Jewish physics."

Willi Menzel's role in the concerted campaign of character assassination against Heisenberg is significant because the National Socialists were most concerned with winning over German youth. One of the most effective methods for grabbing and holding the attention of university students were mandatory political reeducation camps, often devoted to specific topics within the context of the National Socialist "People's Community": the new national, and racially homogeneous community which would eliminate class distinctions and social inequality. This community was often more propaganda than reality, but many Germans had to make at least symbolic gestures towards a classless society. University professors were pressured to attend indoctrination camps where they would mingle with Germans from all classes and professions. If a young scientist wanted to get a teaching job or perhaps a promotion, then in practice he was forced to attend a similar camp as well.

In early 1936 a "physics camp" was held at Darmstadt for university students from throughout the Reich. The teaching staff was dominated by four adherents of *Deutsche Physik*, all of whom had received teaching positions during the first years of the Third

Reich: Alfons Bühl, professor at the Technical University at Karlsruhe, Prof. August Becker, Lenard's successor at the University of Heidelberg, Rudolf Tomaschek, professor at the Munich Technical University, and Prof. Ludwig Wesch, also at the University of Heidelberg. They were joined by three other physicists, including Dr. Wilhelm Dames from the Education Ministry.[95] Menzel was one of the students attending the camp. He wrote the official report on the camp's accomplishments, and sent a copy to Stark.

Alfons Bühl told the assembled physics students that physicists had gotten a bad reputation because they had not paid enough attention to practical matters. Physics had to be made relevant for society at large. The training of science teachers was fundamentally wrong: teachers knew the laws of quantum physics and wave mechanics, but little of applied and experimental physics. The influence of Jewry had made the physicist into a desk physicist. Perhaps most important for the students, Bühl argued, was the historical study of physics through Lenard's *Deutsche Physik*, including examinations of the influence exerted by Catholicism and Jewry, as well as the worldview of "Nordic" physics.

The adherents of *Deutsche Physik* did not forget to attack "Jewish physics." Science had been greatly affected by the influence of Jewry since the end of the first world war, they claimed. Jewish research was little more than mathematical formulas. The theory of relativity was mental acrobatics. While the "Aryan" physicist drew pleasure from nature, the Jewish physicist relied on self-made formulas. Mathematics was merely an auxiliary aid. Finally, Bühl brought up Heisenberg in this context: he possessed a mathematical, constructive, and "Jewish" mind.

Dames, who represented REM and was neutral on the subject of *Deutsche Physik*, argued that a physicist had three tasks: long-term research; immediate applications—for example, the use of physicists in World War I; and political and ideological work. Pure science was insufficient, rather applications were required. When Heisenberg's name was mentioned, a student from Leipzig said that the physicist was a genius. Dames replied that Heisenberg was interested only in pure science and therefore was seen as a genius.

But in careful contrast to Bühl, Dames allowed that one day Heisenberg might abandon his one-sidedness and appreciate practice. Dames took care to echo National Socialist ideas even while distancing himself from the specific doctrines of *Deutsche Physik*. The National Socialist ideology of physics was based upon militarism and racial solidarity.[96] This physics camp is important because it makes clear that the *Deutsche Physik* of Lenard and Stark had no monopoly on "Nazi physics." The Third Reich was interested in science that would help further their long-term goals of racial purity and military expansion. As Dames made clear, even Heisenberg would be acceptable, if the National Socialist state found his physics valuable.

Although Stark's career and the fight against "Jewish physics" appeared to be going well, his attention was diverted by a serious political threat from Adolf Wagner, one of the most ruthless and powerful of the National Socialist regional party leaders. Stark became embroiled in local party politics and challenged Wagner's authority by accusing a local party leader, Endrös, and a local mayor, Karl Sollinger, of improper conduct and damaging the prestige of the NSDAP.

In early 1934 Stark told Wagner that the Endrös matter was so important that Stark felt obligated to make a formal written complaint. Endrös had misused his position as local party leader to intervene illegally in a financial matter and thereby shield an acquaintance who had defrauded both the local government and a widow. Such a man should at least be removed at once from his Party offices. Moreover, since Endrös used lies and slander against his enemies, Stark assumed that he was also using them against him.[97] Nothing happened to Endrös, but this matter was just the beginning of Stark's struggle with the party officials in Stark's home town of Traunstein and the surrounding region of Upper Bavaria.

Less than a year later Stark intervened again in the local politics of Wagner's region, with serious consequences. Karl

Sollinger, Traunstein mayor and city leader of the NSDAP, had been arrested on the authority of Justice Minister, Franz Gürtner, who significantly was not a member of the NSDAP but rather was one of the many representatives of the old order who had helped Hitler into power and who shared power with the National Socialist movement during the first years of the Third Reich. Wagner contacted Gürtner immediately. Although Wagner admitted that the offenses of Sollinger and comrades should not be condoned, they should merely be warned. The state had no interest in the carrying-out of his sentence, since the desired goal could be achieved merely by announcing and suspending the sentence.[98]

Gürtner responded by going over Sollinger's offenses in detail. Sollinger had been sentenced by the special court in Munich in October 1934 to eight months imprisonment for resisting the state's authority and causing dangerous bodily harm. On 20 August 1934, when police commissioner Betz announced the curfew in the local tavern, Sollinger refused to go home. Betz was then brutally beaten and stabbed by Sollinger and others. Wagner advised Sollinger to ignore the sentence. When the party leader told Gürtner that the sentence could not be carried out at that time for reasons of state and party, the Justice Minister agreed.

Sollinger was subsequently sentenced again by a Traunstein court to six months prison and 50 Reich Marks penalty for embezzling from the Winter Relief Fund. This fund was a supposedly voluntary collection taken up by the National Socialist movement, but in fact was a type of coercive tax designed to raise funds and force people into making public shows of support for the National Socialist cause. Fortunately for Sollinger, this sentence was eliminated in the general pardon decreed by Hitler on 7 August 1934— but his guilt remained clear.

Sollinger's conduct and his apparently successful attempts to avoid punishment had caused considerable unrest in the area of Traunstein. This had gone so far that Stark, who owned an estate in the Traunstein area and was considered a party spokesman by the local population, had repeatedly come to Gürtner and argued that it was an urgent necessity in the interests of state and party that Sollinger's sentence be carried out. Gürtner had nevertheless been willing to let Sollinger go unpunished, but the latter finally forced Gürtner's hand. Stark informed the Justice Minister that Sollinger had once again clashed with the police by refusing to obey the curfew. Moreover, Sollinger had bragged about his power, claiming that he would never obey the police, and that his friend Wagner would always protect him. Worst of all, Wagner had hushed up this incident.[99]

Once Wagner's staff knew that their party comrade Stark had denounced Sollinger, they began a concerted campaign of character assassination. First, they told Hitler's personal chancellery that although Stark did have the confidence of a portion of the local population, these were the people who were hostile to National Socialism. Stark wanted to shake up the Traunstein leadership merely because the local leader had once alienated him. In any case, Stark did not have the right to interfere in party political matters. He could not judge whether or not the punishment of Sollinger was in the interest of the state or party. This decision could only be made by the responsible party and state authorities. Stark knew very well, Wagner's staff added, that both the local and regional authorities had always backed Sollinger.[100]

Wagner's own reaction was swift and severe. He began legal proceedings to throw Stark out of the NSDAP. If Stark wanted to complain about the conduct of a party comrade, then he should have made his report to his regional leader. Moreover, Stark had known that Wagner had taken Sollinger's side against the Justice Ministry. By taking a party matter to the Ministry of Justice—which was not controlled by a National Socialist—Stark had caused considerable public damage to the image of both Wagner and the party.

Wagner provided a cynical and hypocritical justification for the process against Stark: a party comrade should not treat another party comrade badly or damage the image of the party.[101] When Wagner's staff submitted their application for Stark's expulsion to the Berlin party court, they also referred to a February 1936 decree by Rudolf Hess, Hitler's personal representative for party affairs and the highest ranking official in the NSDAP: every party comrade who filed complaints in party matters to external state authorities would be expelled.[102]

The Bavarian Party leader then contemptuously told off Gürtner. Wagner had asked Hitler for a pardon for Sollinger, whom Gürtner had imprisoned due to Stark's intrigues. The Justice Minister had no idea of the damage he had done to Hitler's political movement and the National Socialist state. Now action had been taken to throw Stark out of the NSDAP for imprisoning a party comrade by denunciation. Moreover, there was no doubt in Wagner's mind what the outcome of this process would be. Stark's days in the party were numbered.[103]

Stark's first reaction to his threatened expulsion was to demand that the court secure and examine the files from the previous court cases he had brought against Endrös and the counter-suit Endrös had brought in turn, as well as the Sollinger records. Stark suspected that these documents would reveal that Wagner's representative Nippold had intervened illegally on Endrös' side.[104] A few days later Stark went further and applied for Wagner's expulsion from the NSDAP, an extremely unlikely outcome which either demonstrated Stark's fearlessness, his rage, or his naiveté.

The physicist accused Wagner of vile defamation of character and damaging the prestige of National Socialism in the Sollinger case. Wagner had told the regional court in Upper Bavaria that Stark had already been thrown out of the NSDAP by the party leadership. The same claim was disseminated in the region of Traunstein by local party officials. Wagner's obviously untrue claim had defamed Stark's character in Traunstein. Moreover, this internal party matter spilled over to Stark's professional reputation. Wagner had also spread this falsehood in REM and thereby questioned Stark's character within the ministry. Indeed Wagner's slander had even became known among Stark's employees at the PTR. This character defamation was especially incriminating for Wagner because he knew that Stark had publicly supported Hitler as early as 1924 and had worked hard for the National Socialist movement during the last years of the Weimar Republic.

Stark reminded the court that he had held many large public rallies for the National Socialists near Traunstein. He thereby won the confidence of many people in the region for himself and the National Socialist movement. Therefore, Stark felt responsible for seeing that NSDAP functionaries were held to the fundamental principles of National Socialism, for which he had fought. In particular, Stark had certainly done more for National Socialism than either Endrös or Sollinger. Since Stark had gone to Wagner twice with no result, the latter had no right to be upset that the physicist did not go to him a third time. Stark had always acted loyally and correctly, while Wagner had failed in his duty by doing nothing. Even though Sollinger had almost killed the policeman, Stark emphasized, Wagner immediately freed him from jail.[105]

The party court in Berlin examined the Stark case, but told the highest party court in Munich that a trial against Stark appeared unjustified. How could the party completely back one political leader, who had been found guilty by state courts, and simultaneously expel another party comrade, who from a party standpoint had not gone through the proper channels and thereby acted incorrectly, but at least had acted with a clear conscience?[106] The Munich court decided to handle the Stark matter itself.[107] Martin Bormann, Hess' second-in-command, now took a personal interest in the Stark case, most likely because of the physicist's standing as one of Hitler's earliest supporters.[108]

The conflict with Wagner and looming expulsion from the NSDAP made Stark vulnerable. In February 1936, Rust told Hitler that Stark, who was already overwhelmed by his two presidencies, had also offered his services as president of the KWG.[109] Stark fought back by telling Lenard that Rust was a liar. Stark would not become Planck's successor even if he was asked. Rust clearly found it useful to portray Stark as power-hungry[110] and certainly did not want to see him become president of the KWG.

Stark had now fought for nearly two years against what he considered the criminal intentions of Rust's subordinates in the hope that the minister would finally come to his senses. But Stark's patience had come to an end, he wrote Lenard on 11 April. If his desired changes were not made by the end of the month, then he would ask Hitler's permission to retire from his two presidencies. Under the present circumstances, Stark said, his work had been made impossible.[111] Lenard asked Stark to wait at least until the presidency of the KWG had been decided.[112]

But on 29 April, Stark wrote him that the situation had now deteriorated. Rudolf Mentzel, an influential bureaucrat in the Ministry of Education who, in Stark's words, was young, narrow-minded, unscrupulous, and power-hungry, enlisted Vahlen's assistance to cut the Research Foundation budget from 4.7 to 2 million Reich Marks. Furthermore, Mentzel retained power over 1 million of that, and would transfer the remaining million to Stark only on a case-by-case basis, each time requiring Stark to seek Mentzel's approval. Stark had now been made superfluous and felt that the only honorable thing for both him and German science was to resign. Any appeal to Rust would be pointless.[113]

Stark had a knack for making enemies, both within the scientific community and the National Socialist movement. As if he did not have enough problems, in the following months he managed to alienate the Ahnenerbe, the scientific research branch of the SS. Stark denied the SS research DFG funds because he did not consider their projects scholarly enough.[114] The subsequent internal SS report to chief Heinrich Himmler spelled out the problem. Although Stark was a National Socialist, the SS official noted that he did not have the slightest comprehension of politics within the National Socialist movement.

Unfortunately for the SS, Stark believed that science should serve the National Socialist state, but was nevertheless an objective search for truth pursued according to international standards. In other words, what was good science would be determined by the international scientific community according to traditional requirements for research and publication. In Stark's mind there was no contradiction between this stance and his *Deutsche Physik*.

The SS took the position that science, like everything else in the Third Reich, should obey the National Socialist leadership and be determined by the requirements of politics and ideology. Good science was research that provided Himmler with the results he wanted and needed. Thus when the Ahnenerbe complained that Stark did not have the slightest understanding for those sciences which had been reinvigorated during the last three years by National Socialism, it was in fact referring to the physicist's rejection of pseudo-science designed to serve National Socialist ideology and policy. Stark had no problem with the ideology or policy, but he refused to fund pseudoscience with funds from the German Research Foundation.

The SS feared that if the combined pressure of Himmler and Rust could not make Stark and the DFG appreciate the work of the Ahnenerbe, then the SS would have to finance the research by itself.[115] In fact there was a third solution: force Stark to resign. The physicist had never had the support of the scientific community for his presidency, had alienated REM and the SS, and was fighting to stay in the NSDAP. Mentzel had effectively reduced the DFG president to a figurehead. All that remained was an excuse to push Stark out to pasture, for despite what Stark had told Lenard, he now clung to power.

The opportunity came when one of Stark's funding decisions blew up in his face. He invested considerable sums of Research Foundation money in order to subsidize a scheme to refine gold from peat, but the process was worthless and the peat bogs had no gold.[116] Stark was forced to resign by the threat of a public scandal. REM offered him a deal: if he resigned from the DFG, then he could keep the presidency of the PTR. Mentzel, one of Rust's most powerful aides and an honorary SS member willing to support the Ahnenerbe research, was his successor.[117]

As usual, Stark did not hide his frustration from Lenard, his comrade-in-arms. Now that he was rid of the heavy burden of the DFG presidency, he wrote Lenard in November of 1936, he felt psychically and physically relieved and was pleased to be able to devote himself more to scientific work. For two and a half years Stark had fought as president of the DFG, not only for German science, but also against what he called its bureaucratization.[118]

In other words, Stark saw himself as having been fighting almost single-handedly for a second revolution in German science which would go far beyond the initial National Socialist purge of the civil service. His real opponents were not the "friends of the Jews," rather the National Socialists now running the state bureaucracy. But by now the leadership of the Third Reich had little tolerance for such uncoordinated, unsolicited, and unwelcome agitation. In the summer of 1934 Hitler had used the SS to purge the SA (*Sturmabteilung*, translated as Stormtroopers) leadership in the "Night of the Long Knives," murdering Ernst Röhm and other officials who had threatened Hitler's position by their persistent calls for a far-reaching second National Socialist revolution.[119]

Stark did have allies and sympathizers who offered their solace. A member of Hess' staff was shocked by the news of Stark's resignation and asked for the details so that he could pass them on to his boss.[120] Another letter of condolence cast some light on Stark's mismanagement of the Research Foundation. Although a colleague from Alfred Rosenberg's party office was personally moved by the news of Stark's resignation, he was very surprised by the form which the physicist chose for expressing his thanks: a check from the DFG account. Since the Rosenberg official was already compensated for his work in Rosenberg's office and the DFG funds were limited, he returned the check.[121] Not everyone turned down Stark's offer. A staff member at Hans Frank's ministry noted Stark's resignation from the DFG with sincere regrets and great concern. The check the physicist had sent him was further proof of his great generosity.[122]

Stark's successor Rudolf Mentzel was not pleased by the physicist's last minute generosity with DFG money and subsequent threat to cut off all cooperation between the PTR and DFG unless Mentzel provided the PTR with additional funds. Mentzel replied that, since Stark had left him 1.8 million Reich Marks in commitments but only 1.5 million in the bank, it would not be possible to spend more money anytime soon.[123] Stark softened his tone and assured Mentzel that, if he could count on the understanding and cooperation of the Research Foundation in the future, then he was prepared to support the DFG.[124] The PTR president even went so far as to make the token gesture of transferring 3000 Reich Marks from his special president's fund back to the DFG. Mentzel welcomed the transfer as evidence of Stark's willingness to cooperate.[125]

By the time of this last exchange in February 1937, the party court proceedings against Stark were already underway.[126] Stark testified that he had gone to the Reich Ministry of Justice with the Sollinger case out of concern for the prestige of the party and state. It was in their interest that Sollinger serve at least a token sentence. Shortly thereafter Stark had visited Hans Frank, a leading National Socialist lawyer, and said the same thing. Stark had spoken once with Wagner and twice with Endrös on this matter, as well as sending Wagner a letter. When Stark went to the Justice Ministry, he had been unaware that he was going against Wagner's will, although this became clear later.

In short, Stark denied that his discussion with the Justice Ministry was in any way undisciplined. He had been doing a service for both the party and the state. If Stark had known that all such complaints should have gone through Hess in his function as Hitler's personal representative for party matters, then Stark would have done so.

But Stark's real and most effective defense was his long-standing and valuable service to Hitler's movement. As he reminded the court, in 1932 and early 1933, the physicist had made countless political campaigns in Traunstein and the surrounding region for the NSDAP and thereby gained prestige as a spokesman for National Socialism. When the glaring injustice of the Sollinger case took place, Stark believed that he was obligated to ensure that this matter would be handled in a way which corresponded to the interests of the party and the state.[127] The Sollinger case threatened to expose a double standard: party comrades and non-party comrades were being treated differently. Finally, Stark took care to tell the court once again that Wagner had spread lies about him and demanded an expulsion process against the regional leader.[128]

The court could not tell after hearing Stark's testimony whether the physicist had consciously gone against Wagner's will, or had been proceeding with a clear conscience.[129] Wagner in turn angrily denied the scientist's claim of ignorance. Although Stark had repeatedly pushed the Party leader to do something about Sollinger, Wagner had always refused. But that was not the point. Even if Stark's claim had been true, Wagner insisted that, as a long-standing National Socialist, the physicist should have known that a National Socialist did not sell out a party comrade to the Ministry of Justice. However, Wagner now saw fit to be forgiving. Since Stark had fortunately lost the presidency of the DFG, he had been punished enough. Wagner was prepared to halt the expulsion process, so long as Stark recognized his error and apologized to both Wagner and Sollinger in writing.[130]

Johannes Stark began the Third Reich with a great deal of political influence, perhaps more than any other German scientist. But he had already squandered most of his power by 1937, before he made his famous public attack on Heisenberg. Thus this attack was not the result of Stark's success in the Third Reich, rather of his failure.

The Alienation of an Old Fighter

The "White Jew" Stark's situation in the summer of 1937 was grim. He had been forced to resign from the DFG after years of struggle with party comrades in REM. Since Stark refused to humiliate himself by apologizing to Wagner, his case before the highest party court threatened to throw him out of the NSDAP. After having lost so much already in the Third Reich, Stark decided to fight for what he had left—the purity of *Deutsche Physik*. Once again, Stark adopted a strategy of character defamation in order to deny the Munich professorship to Werner Heisenberg, but this time Stark took the consequential step of allying himself with forces within the SS.

By the middle thirties Stark had become contemptuous of the "dogmatic" theoreticians of his time, who he claimed were no longer capable of understanding experimental physics.[131] Such theoretical physicists produced work which conflicted with reality and remained silent about uncomfortable facts.[132] But it was the combination of Stark's long-standing feud with Arnold Sommerfeld, the fact that Munich lay in his native region of Bavaria and was the capital of the National Socialist movement, where Hitler's movement had gotten its start, and Stark's recent setbacks that pushed him beyond his previous ideological excesses and led to vicious and dangerous personal attacks on Heisenberg. If he could not defeat his party enemies, he could at least try to gain some satisfaction in the fight for the ideological purity of physics.

In February 1937 the Bavarian Ministry of Culture requested that Heisenberg be called to the Munich professorship.[133] But the head of the Reich University Students League appealed Heisenberg's appointment. Ludwig Wesch hoped that if Heisenberg could be kept out and the call of the *Deutsche Physik* adherent Rudolf Tomaschek to the Munich Technical University went through (as it subsequently did), then there would at least be one stronghold of "Nordic research" standing guard in Munich.[134]

Stark was now forced to ask the hated REM for assistance. He called Dames in June 1937 concerning the Munich appointment, but he was told that as an outsider he could not be granted access to the files or the candidate list submitted by the Munich faculty. However, REM would be pleased to hear Stark's suggestion for the post.[135] Stark now took a step designed to force REM's hand and keep Heisenberg out of one of the few *Deutsche Physik* strongholds: he used the SS to attack Heisenberg's character.

On 15 July 1937 an anonymous article appeared in the SS weekly *Das Schwarze Korps* (literally translated as "The Black Corps") shamelessly[136] attacking Heisenberg as a "white Jew" and the "Ossietzky of physics."[137] The chilling term "white Jew" described an "Aryan" who had been tainted or contaminated by Jewish spirit. The equally threatening label "Ossietzky of physics" referred to the socialist and pacifist Carl Ossietzky, who had provoked Hitler's rage by receiving the Nobel Peace Prize while imprisoned in a concentration camp—where he died.[138] Such personal attacks were exceptionally dangerous for the individual target, but in the long run proved ineffective as far as official policy towards physics was concerned.[139]

Friedrich Hund, a colleague of Heisenberg at the University of Leipzig, told the rector that the purpose of the *Schwarze Korps* article was clearly to hinder Heisenberg's call to Munich.[140] Stark's own contribution, "'Science' Has Failed Politically," immediately followed "White Jews in Science" and made clear who was behind the attack. Stark pointed out that German science had manifestly failed to rally to Hitler's cause. Even though the Jews were gone, Stark cautioned that most of the Jews' "Aryan" comrades and students remained in their positions. Finally, Stark dismissed arguments that these scientists were indispensable for the economy and national defense.[141]

In fact, the physicist had not overcome the hostility the SS had for him. Somehow Stark had gained only the assistance of the rather independent editor of *Das Schwarze Korps*. But it certainly appeared to the general public that the SS had now thrown its weight behind *Deutsche Physik*. Several leading British scientists brought the article in *Das Schwarze Korps* and in particular Stark's remarks on "White Jews in Science" to the attention of the editor of *Nature*, who wrote Stark on 11 October that he hesitated to make any reference to this report without confirmation that it accurately represented Stark's considered opinion upon the subject of "White Jews." The scientific world, the *Nature* editor added, would be interested in knowing Stark's views on the "relation of a certain group of people to scientific progress."[142]

Stark was flattered by this request and immediately replied that he would be pleased to provide *Nature* with an article on the influence of Jews in German science.[143] *Nature* responded quickly in turn and requested an article of 1,000 to 1,500 words on the subject of "Jewish influence on science in Germany or elsewhere." The editor assured Stark that he was completely independent of either Jewish or anti-Jewish influence, and only desired to promote international cooperation in pursuit of the principles of truth and the progress of natural knowledge.[144] *Nature* may have chosen to contact Stark before publishing any criticism of the articles in *Das*

Schwarze Korps because its editor feared that his journal might be banned in Germany. Indeed in late 1937 *Nature* was proscribed in German libraries[145] after it had been attacked as an atrocity journal.[146]

Stark proceeded cautiously. After his manuscript was finished, he sent it first to a party comrade and high-ranking official in the Ministry of Propaganda for approval. Stark told him that he had been leading a tough and bitter struggle against the "Jewish spirit" in science. It was very important to Stark that Heisenberg, who he called the champion of Jewish influence, not be honored with a call to the university in Munich. This goal had been served by the article which appeared in *Das Schwarze Korps* and which had incited international Jewry against Stark even more than before. Jews and their comrades were now attacking Stark in *Nature*, a journal with a world-wide distribution. Fortunately, its editor had been decent enough to contact Stark. The enclosed article had been written with scientific objectivity and in Stark's own words was pitched to the Anglo-Saxon and "non-Aryan" psyche. Of course, Stark hastened to add, when he wrote other publications for Germans, he naturally was clearer and more concrete.[147]

Stark's article, "The Pragmatic and the Dogmatic Spirit in Physics,"[148] provides a good opportunity to examine his often tortured arguments concerning "Aryan" and "Jewish physics." In his *Nature* article he could not simply use National Socialist slogans and threats in order to silence opposition, but rather had to limit himself as much as possible to rational argument and logical persuasion. Stark admitted that physical science itself is international, that is, the laws of nature are independent of human existence, action, and thought, and are the same all over the world. However, he insisted that the manner in which physical research is carried out depended on the spirit and character of the scientists involved.

There were two principal types of mentality in physics, the "pragmatic" and the "dogmatic." The pragmatic scientist wants to discover natural laws by means of experiment. He may use theoretical conceptions, but if they do not agree with the experimental results, then the theory is abandoned. The pragmatic goal is to establish reality. In contrast, the dogmatic scientist begins with a theoretical conception based on ideas he has created, uses mathematics to elaborate them, and finally seeks to give them physical meaning.

If they agree with experimental results, then the dogmatic scientist emphasizes this agreement and implies that these experimental results could only have been established and only have scientific importance because of his theory. But if the experimental results do not support his theory, then he questions their validity or considers them so unimportant that he does not even mention them. Furthermore, Stark claimed that dogmatic physicists imply that their theories and formulas cover the whole range of phenomena. They do not see any further problems in this field, rather their formulas freeze any further thought or inquiry.

According to Stark, this difference between pragmatic and dogmatic physics has important consequences. Whereas the "pragmatic spirit" leads to new discoveries and knowledge, the "dogmatic spirit" cripples experimental research and is comparable to the theological dogmatism of the Middle Ages. Stark then put faces to these labels. The German experimental physicist Philipp Lenard and his British counterpart Ernest Rutherford were pragmatic. Both had made important experimental discoveries, the former for the connection between the electron and light, the latter in radioactivity and the nuclear structure of atoms. In contrast, Stark labeled the theoretical physicists Max Born, Pascual Jordan, Werner Heisenberg, Erwin Schrödinger, Arnold Sommerfeld, and more importantly, Albert Einstein, dogmatic. Their work was arbitrary and "physical-mathematical acrobatics."

But what disturbed Stark the most was not the dogmatic theories themselves, rather *how* they had become influential,[149] the same criticism he had made in 1922. The pragmatic physicist did not conduct propaganda for his research results. But the protagonists of the dogmatic spirit were very different. They did not wait to see whether or not their theories might prove to be inadequate or incorrect. Instead they use articles in journals and newspapers, textbooks, and lecture tours to start a flood of international propaganda for their theories, sometimes almost before they have even been published. Neither Lenard nor Rutherford used lecture tours to promote their results, Stark noted, but propaganda for Einstein's theory of relativity had been carried to a wide public around the world.

Stark now turned to the specific situation in Germany. During the previous three decades the representatives of the dogmatic spirit had become dominating with the help of the governmental bureaucracy, in particular by acquiring many physics professorships. This domination of academic physics, together with lively propaganda for modern dogmatic theories, meant that much of German academic youth was educated in the dogmatic spirit. Stark had repeatedly observed the crippling and damaging effect this domination had had on the development of physical research in Germany.

Finally, Stark turned to the matter of the Jews. He had opposed the damaging influence of Jews in German science because they were the chief exponents and propagandists of the dogmatic spirit. According to Stark, the history of physics demonstrated that the founders of physics research, and the great discoverers from Galileo and Newton to the physical pioneers of his own time were almost exclusively "Aryans," "predominantly of the Nordic race." Thus Stark concluded that men of the "Nordic" race were predisposed towards pragmatic thinking. In contrast, the originators, representatives, and propagandists of modern dogmatic theories were predominantly "men of Jewish descent." Moreover, Jews had played a decisive part in the foundation of theological dogmatism and were mainly responsible for Marxism and communism. Thus Jews were naturally inclined to dogmatic thought.

Stark finished his article with several qualifications. Of course there were "Aryan" scientists who were dogmatic, and there were Jews who could produce valuable experimental work in the pragmatic spirit. "Aryans" could become accustomed by training and practice to dogmatism and Jews to pragmatism. Stark would welcome scientific achievement and new discoveries no matter who made them. He combated the harmful influence of the dogmatic spirit in physics whenever he encountered it, whether the culprit was a Jew or not. Moreover, Stark noted that he had been fighting this battle since 1922, not 1933.

In other words, Stark's juxtaposition of pragmatic and dogmatic physics had two complementary sides: (1) an experimental physicist's rejection and lack of appreciation of modern theoretical physics, compounded by his own personal and professional bitterness; (2) his own personal brand of anti-Semitism and support of National Socialism. Stark thereby rejected the two most common National Socialist attitudes to physics (or indeed to science): either (1) an opportunistic approach, whereby if scientists and science were useful for the state, then they would be used; or (2) an idealistic approach, whereby a Jewish scientist was a Jew first—and therefore an enemy of Germany—and scientist second. Since Stark fell in neither camp, he could be sure of support from neither.

When his comrade-in-arms Lenard criticized Stark for publishing in what he called the "Jewish journal" *Nature*, Stark's growing alienation and bitterness became crystal clear. Stark's struggle against the "Jewish spirit" had been systematically boycotted by the influential German authorities. Indeed influential forces in the National Socialist state had begun to forsake him and instead either line up behind scientists like Heisenberg or remain neutral. In 1936 Alfred Rosenberg stopped taking Stark's articles in the *Völkischer Beobachter* and in Stark's opinion had become "the protector of the friends of the Jews." *Das Schwarze Korps* no longer accepted Stark's articles as well. The SS began an investigation of Heisenberg immediately after the 1937 article attacking "white Jews," which ended with Heisenberg's political rehabilitation. Under these conditions Stark had to be grateful to the editors of *Nature* for the invitation to bring the influence of Jews and the Jewish spirit before a large international public.[150]

Ironically, the articles in *Das Schwarze Korps* and *Nature* convinced very many people inside and outside of Germany that Stark was very powerful indeed, perhaps even the dictator of physics he claimed. In fact, when the *Nature* article was published in the spring of 1938 Stark's influence had peaked and was fading fast. In particular, the main result of the article in *Das Schwarze Korps* was that the head of the SS, Heinrich Himmler, threw his support behind Heisenberg and forbade any further attack.[151] As the head of the SS explained to his subordinate, Germany could not afford to lose Heisenberg, who was relatively young and could train another generation of scientists,[152] something Stark and Lenard obviously could not do.

Since Stark had refused to apologize to Wagner, the supreme party court scheduled his trial to begin in the fall of 1937.[153] Stark's trial had been repeatedly delayed because the court records of the relevant previous trials in Bavaria had not arrived. Wagner had apparently hindered their transmission in the hope that Stark's case would be decided without them. When the Highest Party Court made clear that they would not proceed before they arrived, the local court officials in Wagner's region finally relinquished them.[154]

Stark described this trial as the tragic end of his fourteen year struggle for Hitler and his movement[155] and flatly rejected the charges against him. Appealing to the Justice Ministry was no offense against the NSDAP and the fact that a regional leader disagreed did not make it so. Stark was not responsible to Wagner and the latter's opinion was hardly identical to that of the NSDAP. Indeed Wagner had demonstrated through his conduct that he, not Stark, did not deserve to belong to the NSDAP.

The PTR president and old fighter was shaken by the fact that the Highest Party Court began a trial against him for conduct which he had felt obligated to do precisely in the interest of the party. Stark had been fighting longer for Hitler and the NSDAP than had Wagner, and could judge for himself what benefited or damaged the party. Moreover, the physicist had no intention of taking his expulsion quietly: he would inform Hitler personally of the tragic end of Stark's struggle for the NSDAP and its *Führer* (Hitler's title, literally translated as "leader"). Hitler, Stark was convinced, would not judge his conduct as an offense against the efforts of the party.

Once again, the physicist went down the list of his distinguished service to National Socialism. Stark began supporting Hitler publicly in 1923 and in particular when the National Socialist leader was imprisoned after the failed Beer Hall Putsch. In 1930 Stark sacrificed his scientific work in order to help put Hitler and the National Socialists over the top. The *Führer* subsequently thanked the physicist heartily in the name of the party for his work. Even after the National Socialists came to power, Stark continued to fight for Hitler and National Socialism, for example in his *Nature* articles. Scientists outside of Germany, Stark claimed, considered him both the most respected and most hated "Nazi Professor."

Lately Stark had been fighting within Germany against the scientific influence of Jews and their comrades. This struggle had led to a cowardly conspiracy against him, whereby influential party comrades harassed Stark and tried to stain his reputation. Wagner's efforts against him were all the more bitter because the Party leader had been Stark's student in Aachen where the professor had benevolently assessed Wagner's examination, i.e., had given Wagner a grade he really did not deserve.[156] Finally, when Stark came to the end of his statement, he did not merely ask to remain in the party. He demanded again that the court give him satisfaction and expel Wagner.[157]

After careful consideration of all the testimony and evidence, the Munich court saw no point in proceeding with Stark's trial. There was no doubt that Stark truly believed that Sollinger should have been disciplined. Stark could be punished only for not going through official party channels to Hess with his complaints. Since Sollinger had not been punished in any way—and obviously would not be—and Stark had already lost the presidency of the Research Foundation, the court intended to stop the proceedings—if Hess agreed.[158]

The NSDAP leadership agreed that the trial should be quashed. Indeed, Hess' office remained one of the few forces within the National Socialist state that continued to support Stark, possibly because he was an old fighter.[159] Although the physicist should have taken his complaint to Hess, the court had to agree with Stark that the Sollinger affair had hurt the image of the party. Stark may also not have known that he should have gone through Hess. Thus he had very little guilt. The court decreed that no punishment was necessary, especially since the accused had performed valuable services to the National Socialist movement during the "time of struggle," as the National Socialists described the Weimar Republic.[160] Stark could now stay in the party, even though he had already become an outsider. In many respects the struggle with Wagner left him a broken man.

All that Stark had left was the fight to deny the Munich professorship to the "dogmatic" "white Jew," Heisenberg. In the end the public attack in *Das Schwarze Korps*, together with the steadfast opposition of Hess' office, killed the appointment despite Himmler's support of Heisenberg. The main party office first rejected Heisenberg, then argued that it could not change its mind for reasons of prestige. REM had previously offered the job to Heisenberg, but now fell in line behind the Party Chancellery. Even Himmler was only willing to promise Heisenberg a prestigious appointment somewhere other than Munich.[161] Heisenberg and Sommerfeld had little choice but to acquiesce.

But who would succeed Sommerfeld? In early 1938 Stark asked Bruno Thüring, astronomer and *Deutsche Physik* adherent, to take over the professorship for theoretical physics temporarily. If all went well, he might be able to succeed Sommerfeld. Stark was not worried by the fact that Thüring was not a theoretical physicist. Indeed Stark argued that it would be easy for his younger colleague to give reasonable, not too detailed lectures on theoretical physics. Most importantly, Thüring would bring a new spirit into the Munich faculty. If he was interested, then Stark would suggest him to REM.[162]

Thüring discussed Stark's suggestion with the local National Socialist officials in Munich and replied that, for political reasons, he was prepared in principle to take over the professorship temporarily as a last resort. However, he had more professional scruples than Stark and was unwilling to take the job permanently. He was an astronomer, not a theoretical physicist. Moreover, it was well known that Thüring was already involved in the fight to keep Heisenberg out of Munich. If Thüring would now take the job, then he feared that his future career would be tainted with the stigma of a cold-blooded careerist, which would not help their fight against "Jewish physics."[163]

The Munich position finally went to Wilhelm Müller, another supporter of *Deutsche Physik*. Stark had been very influential in Müller's career during the Third Reich. In 1934 Stark threw his support behind Müller's appointment at the Technical University of Aachen.[164] Less than a week after the article in *Das Schwarze Korps*, Stark confidentially asked an Aachen colleague about Müller, whom he intended to recommend for a professorship.[165] Müller was eager and willing to join the fight against Einstein and "Jewish physics."[166]

After a long and Byzantine bureaucratic conflict between the Party Chancellery, REM, the University of Munich, and supporters of *Deutsche Physik*, Müller succeeded Sommerfeld on 1 December 1939, three months after the start of World War II.[167] Müller's appointment has often been seen as proof of the power and dangerous nature of *Deutsche Physik*. In fact, it was a Pyrrhic victory. By the end of 1939, *Deutsche Physik* occupied six of the eighty-one professorships available in Germany and Austria. Henceforth their numbers would only decline.[168]

The year 1939 was an ambivalent year for Stark. Müller's appointment was his final success, but in the same year Stark retired from the PTR, returned to his estate in Traunstein,[169] and thereby lost the last political or scientific influence he had left in the Third Reich. Stark and his *Deutsche Physik* became less and less relevant for the Third Reich as the war progressed. Even the appointment in 1939 of Wilhelm Führer, a follower of Lenard and Stark, to an influential position in REM only delayed the fall of *Deutsche Physik*. For example, although Führer strenuously opposed the appointment of the astronomer Otto Heckmann in Hamburg, he eventually had to admit defeat and give him the professorship, due in large part to Heckmann's successful efforts to make himself and his science palatable to National Socialism.[170]

The established physics community also launched a counter-attack against *Deutsche Physik*. Meetings between the two sides sponsored by National Socialist officials in Munich in late 1940, and in Seefeld two years later, practically silenced calls for a more "Aryan" physics. The followers of Lenard and Stark who attended were forced to discuss physics rather than politics, with the result that a party agency officially recognized relativity theory and quantum mechanics as acceptable science and embraced neutrality on the issue of modern physics.[171] After the Munich meeting Heisenberg wrote his mentor Sommerfeld and expressed satisfaction with the outcome. Thüring and Müller, the most fanatical advocates of *Deutsche Physik*, had left before the compromise agreement was signed.[172] Rudolf Tomaschek, considered one of Lenard's best students,[173] had already noticed that the wind was changing.[174]

This victory was only possible because Heisenberg and other supporters of modern physics were willing to make the distinction Himmler had required when he backed Heisenberg's political rehabilitation: Einstein had to be separated from his theory of relativity. Sometimes he was attacked as a Jew, sometimes (unfairly) as a plagiarist, and still other times physicists like Heisenberg merely argued that the theory of relativity would eventually have been discovered by someone else.[175]

A few years later, after Heisenberg's political rehabilitation by the SS had sunk in, after he had become a valuable goodwill ambassador for German science outside of Germany,[176] and after his secret work on applied nuclear fission brought him the support of influential figures in the armed forces and Albert Speer's Ministry of Armaments, Heisenberg was given two prestigious appointments: the directorship of the Kaiser Wilhelm Institute for Physics and professor of physics at the University of Berlin. These appointments were widely seen as a victory over *Deutsche Physik*[177] and no doubt perfected Stark's bitterness towards his enemies within the National Socialist leadership.

Müller's appointment in Munich also turned sour, in part because he was not even a physicist, rather an engineer who had taught applied mechanics at Aachen. He had never published in a physics journal.[178] In 1941 the eminent aeronautical engineer Ludwig Prandtl complained to SS leader Himmler, Reich Marshall Hermann Göring, and high-ranking officials in the armed forces that Müller taught only aeronautical and engineering mechanics.

Although students should learn these things, Prandtl argued that they were denied an essential part of a physics education and their necessary education was thereby sabotaged.[179] Müller's weakness in this regard was symptomatic of a fundamental flaw in *Deutsche Physik*: its ideological hostility towards modern science and technology ensured that it could not compete with its rivals when the German state became more interested in economic and military power than ideological purity.[180]

Stark's exchange with Thüring demonstrated that the senior scientist was not really interested in whether or not Sommerfeld's successor was a theoretical physicist, rather only whether he was willing and able to fight the "dogmatic" spirit in German physics. However, Müller's obvious and fundamental incompetence made him a lightning rod for the attacks by the growing forces arrayed against *Deutsche Physik*. At first it appeared that Müller was holding his own, thanks to political backing from local party officials. REM agreed to transform the Munich institute into an institute for theoretical physics and applied mechanics,[181] thereby undercutting the criticism that Müller taught only mechanics. In the spring of 1941 Müller was named dean of the scientific faculty. When Stark congratulated his younger colleague, he noted with pleasure that only a few years ago this faculty was dominated by the "little Jew-descendent Sommerfeld."[182]

The fight against "Jewish physics" continued, with Munich now replacing Heidelberg as the stronghold of *Deutsche Physik*. But local advocates like Müller and Thüring lacked originality and only repeated what Stark and Lenard had already said. In particular, Müller differed from Lenard and Stark only in the violence of his language, describing the theory of relativity as "magical atheism," "pseudophysics," "swindle," "Talmudic inflation-physics," "unscrupulous falsification of reality," and the "great Jewish world-bluff."[183]

However, it soon became clear that Müller did not have the nerve to lead the fight against "Jewish physics," especially when he became the victim of the same sort of tactics *Deutsche Physik* had used against their enemies. Sommerfeld's institute mechanic, Karl Selmayer, remained loyal to Sommerfeld and began to torment Müller, who denounced his mechanic in turn as the tool of the "Jew-comrades" Sommerfeld and Gerlach. Since Selmayer was also an Old Fighter in the NSDAP and enjoyed the support of National Socialist university officials, there was little Müller could do except complain, which he did profusely.[184] By the end of 1941, conditions in Munich had deteriorated so much that Müller threatened to leave Munich if the harassment of him and his co-workers was not stopped.[185]

Müller demanded support from the local party leadership and complained about the rumors which were being used against him. Within a little more than a half a year, emissaries of the university rector pressured Müller to resign as dean. He told the rector that recent events had hit him so hard that he was afraid of a complete nervous breakdown.[186] In the fall of 1942 Müller's complaints to his party allies took on a pathetic tone. From the beginning Müller's appointment in Munich had been a sacrifice which he had accepted freely as a National Socialist because Müller believed that he was serving a holy cause.[187] If personal wishes had been most important, Müller told Stark in 1943, then he would no longer be in Munich.[188] Müller managed to hold out in Munich to the end of the Third Reich, but then ironically was one of the very few scientists to lose his chair through the official postwar Allied policy of denazification and be barred from academia. After the war both Sommerfeld and Selmayer went out of their way to damn Müller before the American Occupation authorities. In contrast, Sommerfeld worked to clear Selmayer's name.[189]

In April 1944 Müller congratulated Stark on his seventieth birthday with the following rather pathetic praise. There were more followers of *Deutsche Physik* than the "dogmatists" wanted to believe. Many independent-thinking engineers and physicists, Müller claimed, were only waiting to be liberated from dogmatism. Unfortunately, the current state of the war hindered the victorious continuation of their struggle, but as Stark had often told Müller himself, it would be rekindled after the war. Müller assured Stark that after their struggle was finally victorious, those men would be remembered who had instinctively carried the flag forward, undaunted by persecution and slander during the early years of struggle and under the harshest "Jewish domination" and who had paved the way towards a future freedom in science.[190]

One of Müller's many problems in Munich was Ludwig Glaser, Stark's former student at Würzburg. Müller immediately hired Glaser as his assistant when he succeeded Sommerfeld, probably at Stark's suggestion. A year previously Stark had asked Glaser to describe the Würzburg events in writing and offered to help Glaser reenter higher education.[191] Glaser's track record as an early opponent of Einstein,[192] the subsequent opposition to his *Habilitation*, and the fact that he joined the NSDAP before the National Socialists came to power should have ensured a successful career under National Socialism.[193] Officials at Hitler's personal chancellery believed that Glaser had a political past in the best sense, was self-confident, tough, and courageous in the service of National Socialism.

During the Weimar Republic, Glaser had restricted his opposition to the theory of relativity to scientific arguments, but he was now more than willing to use virulent anti-Semitic and racist rhetoric in the struggle against "dogmatic" physics. He spoke at eight party functions during his first year in Munich and gave many lectures before groups of the armed forces.[194] His publications during this period were just as enthusiastic:

> The remainder of the Jews, the Jewish half-breeds, and those with Jewish blood have vanished from the academies, libraries, and the lecture halls, and where else they had clung to because of their supposed indispensability ... We thank our leader Adolf Hitler, that he has liberated us from the Jewish plague.[195]

Perhaps more interesting was his apparently unconscious use of National Socialist imagery in an otherwise strictly professional physics article. Glaser described energy quanta as "foreign bodies" in physics. Their "elimination"[196] would be a deliverance.[197]

Unfortunately, Glaser had become too enthusiastic and extreme in almost every way. In June 1941 Bruno Thüring told Müller that Glaser was now a liability to the *Deutsche Physik* movement. He was eccentric. The more his professional prospects improved, the wilder he became. He was an elephant in a china shop. Worst of all, he could not keep his mouth shut. In short, Glaser was a "psychopath."[198] Müller agreed with this judgment and hastened to help Glaser find other employment. Glaser was not saying anything different from Müller or other advocates of *Deutsche Physik*, but he was too much of an idealist to submit to the discipline of either the *Deutsche Physik* movement or the NSDAP.

First, there was an aborted attempt to send Glaser to the reestablished Reich University of Strassburg, a university refounded in what had been French territory as a showcase for National Socialist scholarship.[199] Glaser ended up instead at the eastern counterpart of Strassburg, the Reich University of Posen set up in what had previously been Poland. Glaser was made the provisional director of the institute for applied physics and began a six-part series of lectures on the "Jewish question in science" and the racial nature of science.[200] Ironically Glaser's lectures at Posen demonstrated how bankrupt the idea of a *Deutsche Physik* was. When Glaser, perhaps one of the most extreme followers of Lenard and Stark, finally got an opportunity to teach German youth, he ended up lecturing not on physics, rather on a racist form of history or philosophy of science. There was no uniquely "Aryan" physics which could be taught in a physics course.

Müller soon warned a colleague in Posen to watch Glaser carefully.[201] Müller's assistant had stirred up a lot of trouble for his boss in Munich, but worst of all Glaser had both taken Munich equipment with him to Posen without permission[202] and ordered a wind tunnel—coincidentally from a firm where Glaser's brother was employed and stood to benefit from the deal—without authorization or being able to pay for it. Müller was left holding the bag. When he protested, Glaser reacted by blaming everything on the friends of Jews.[203] Glaser soon wore out his welcome in Posen and had to move on to yet another National Socialist university set up in occupied Europe, the Reich University in Prague. According to postwar records, Glaser disappeared there at the end of the war. Perhaps he died fighting the invading Red Army, a fate befitting a true follower of *Deutsche Physik*.

The failures of Müller and Glaser ruined the only real triumph of *Deutsche Physik*, denying Heisenberg the Munich chair, and brought Stark full circle back to the personal and professional alienation he had felt during the early twenties in Würzburg. Then he had rejected the German republic and his academic colleagues; now he no longer believed in National Socialism and rejected his party comrades. In 1942, when most Germans still believed that Germany could win the war, Stark told Lenard that he was considering leaving the NSDAP because of his struggle with Wagner. Lenard responded with a telegram urging him to reconsider, even though Stark's senior colleague had also been alienated by National Socialism. Hitler, Himmler, and other influential National Socialists listened to the advocates of pseudoscience like the "World Ice Theory," not Nobel laureates like Lenard.[204]

By the end of the war Stark and Lenard had been taught a hard lesson about using political and ideological means to influence science and scientists. National Socialist science policy was a volatile mixture of technocracy and irrational ideology.[205] The technocrats or technocratic institutions in the Third Reich rejected *Deutsche Physik* in favor of science and scientists that were more useful. There were also National Socialist leaders who were unwilling or unable to appreciate high-quality and useful scientists, but such individuals were hardly likely to appreciate even Lenard and Stark. The two senior physicists wanted to have it both ways: to be able to use political and ideological means to attack other scientists, but to have the National Socialist state nevertheless honor, respect, and cherish their own scientific credentials.

There were many instances where Stark did not get his way in the Third Reich, not due to resistance to *Deutsche Physik* within the scientific community, but instead because he was hopelessly outmatched when it came to political in-fighting within the National Socialist state. Stark saw this clearly and early, and knew who to blame. In April 1934 he told Lenard that it would be difficult for he, Stark, to fight for their conception of science and like-minded colleagues. He did not fear the Jews and their other opponents, rather the arrogance, jealousy, and intrigue in the leading National Socialist circles.

They had to see things as they truly were, he emphasized to Lenard. People like Lenard and Stark were not honored by the National Socialist leadership. First, the two physicists were too old and for that reason alone were mediocre. Second, Lenard and Stark had achieved something in their lives, and in the anti-intellectual climate of the Third Reich many of the men around Hitler considered this a disgrace. Third, Hitler was fundamentally unsympathetic towards science. When Lenard and Stark offered their help to the National Socialist leadership, the latter considered the scientists a burden and made sure that Lenard and Stark were aware of their feelings.[206]

The depth of Stark's frustration and bitterness was revealed in the steps he took towards the end of war to leave the National Socialist movement.[207] Stark's son Hans, a National Socialist of even longer-standing than his father,[208] was arrested by the Gestapo for treating a Polish forced laborer too well and then subsequently drafted and sent to the front. When Stark was threatened by local party officials, he and his wife used this as an excuse to submit their resignations from the NSDAP. The matter was referred to the Munich regional leader, who forced Stark to remain in the party by threatening Stark's son.

This sequence of events may subsequently have saved Johannes Stark's life. Towards the very end of the war an SS officer who was quartered at Stark's estate decided that he wanted to keep it. But when he tried to get rid of the Nobel laureate, the local party official refused to support sending such a long-standing party comrade to a concentration camp. At the beginning of May 1945 Stark's house was abandoned by the SS and taken over by representatives of the American military government, who in turn arrested Stark.[209]

✠   ✠   ✠

Postwar After the war the Allies agreed that Germany and Germans should be "demilitarized" and "denazified." All Germans had to fill out a detailed questionnaire on their activities during the Third Reich. A minority of Germans subsequently had to defend themselves in denazification court and risked being convicted of complicity in the crimes of National Socialism. Although the overwhelming majority of German physicists managed to pass through denazification and retain or regain a university position by the early fifties at the latest, the adherents of *Deutsche Physik* were quickly purged from the German universities and kept out.

Since Philipp Lenard, a very old man at the end of the war, died in 1947, Stark had to defend *Deutsche Physik* in denazification court. When the physicist filled out his denazification questionnaire, he argued that he should be cleared of all charges. Instead, the denazification court at Traunstein convicted and sentenced him as a major offender to four years of hard labor. Stark, seventy-three years old and in failing health, appealed.[210]

The Munich court of appeal subsequently reversed the Traunstein judgment. The court broke down the charge against Stark into three parts: conflicts with people in the region of Traunstein; support of Hitler and National Socialism before 1933; and activity as Research Foundation president from 1934 to 1936 and PTR president from 1933 to 1939. The first charge was disposed of quickly, since Stark's accusers were less credible than the accused. The second charge was undeniable, but the Munich court accepted the argument that support of Hitler before the National Socialists came to power was not necessarily support of the subsequent National Socialist dictatorship. Moreover, the court believed

Stark's claim that he had resigned from the party before the end of the war.

The third charge was complicated by the apparently false testimony given in Traunstein that Stark had employed only party comrades as scientists at the PTR. This sweeping claim was revealed to be an exaggeration, although relative to other institutions the PTR may well have had a high percentage of NSDAP members. Furthermore the Munich court heard testimony that Stark had run the PTR in a professionally correct manner.[211]

But the third charge also included Stark's attacks on the supporters of "Jewish science," so the Munich court solicited statements from Einstein, Heisenberg, and others on Stark's anti-Semitism and opposition to the theory of relativity. Ironically the court thereby mirrored the postwar apologia employed by the German physics community. After the war Heisenberg and many other physicists implied that the advocates of *Deutsche Physik* had been the only physicists who had collaborated with the Third Reich and that the collaboration of physics with National Socialism had been limited to the anti-Semitic campaign against Einstein and his theory of relativity.

The followers of Lenard and Stark were anti-Semitic and did oppose relativity, but this in no way constitutes the total perversion of physics by National Socialism. After the war all German physicists were anxious to document their purely academic activities during the National Socialist era and to assert that, by adhering to professional values, they had opposed National Socialism. But such adherence was no opposition.[212] Their activities had not been exclusively academic and their professionalism had merely facilitated greater collaboration with the Third Reich.

Heisenberg was asked two very narrow and specific questions about his conflict with Stark. Was the difference between "dogmatic and pragmatic physics" grounded in anti-Semitism, or in professionally justifiable research methods? Did Stark play a role in the rejection and prohibition of the theory of relativity during the Third Reich?[213] Heisenberg told the court he believed that the attack by Stark on him as a "white Jew" was not due to personal antagonism. Stark had wanted to block Heisenberg's call to Munich.[214] Einstein characterized Stark as paranoid and opportunistic, but not sincerely anti-Semitic.[215] In fact, both Nobel laureates doubted that anti-Semitism had been at the root of Stark's actions. Rather Stark's bitterness at not having been appreciated by his colleagues and government—at least in Stark's mind—had caused what Heisenberg called his preposterous behavior. However, Heisenberg did make clear who was responsible for *Deutsche Physik*. The campaign against the theory of relativity, led by a small National Socialist clique, had been due almost exclusively to the activity of two people. Lenard and Stark, Heisenberg added, had successfully seduced young party members into attacking "senile and Jewishified" physics.

The Munich court of appeals determined that the *Deutsche Physik* controversy was a scientific debate which the court could not judge—ironically the same argument the National Socialist bureaucracy made in 1942, when it rehabilitated Heisenberg—and placed Stark in the group of lesser offenders and fined him 1,000 German Marks.[216] Stark himself went to his grave convinced that he had fought for the freedom of research against REM, that he had only accepted the burden of the Research Foundation presidency in order to forestall its politicization, and that his problems with Wagner proved that he had fought against the injustice of National Socialism.[217]

Thus Stark was able to convince himself that even the very fight for *Deutsche Physik* had been a fight against National Socialism. He was hardly alone. After the war almost all scientists managed to convince themselves (not to mention others) that they had resisted the evil of National Socialism. The eighty-three-year-old Stark died unrepentant in 1957.

✠   ✠   ✠

The Death of Deutsche Physik In his study of scientists under Hitler, the historian Alan Beyerchen argued that the *Deutsche Physik* movement failed because it was neither able to gain backing from political sources nor to win the support of the professional physics community.[218] Lenard, Stark, and their small group of followers remained isolated during the Third Reich and lost what little political influence they had because they were unwilling or unable to serve National Socialism effectively as scientists. Most of the usefulness of *Deutsche Physik* to the National Socialist movement ended when Einstein and the rest of the Jewish physicists had been hounded out of Germany.

For the established physics community under Hitler, a fundamental issue was the extent to which compromise with the regime was necessary in order to retain the greatest possible degree of professional autonomy.[219] But *Deutsche Physik* threatened this autonomy far more than did the National Socialist leadership. Beyerchen notes that the leading figures in the physics community did not seek to embrace National Socialism on its own terms.[220] But neither did Lenard and Stark.

Embracing National Socialism required far more than merely railing against "Jewish physics" and the "friends of the Jews" in science. It also meant a willingness to participate in the cynical politics of the National Socialist state, where principles of any kind had little place, and once the war began, both a willingness and ability to contribute to the German military and economic expansion into Europe and the Soviet Union and thereby to participate in the policies of persecution, exploitation, and genocide.

In the past, emphasis on the "evil Nazi" has often been used—consciously or unconsciously—for apologia, to divert attention from or to deny the responsibility and complicity of the overwhelming majority of German scientists under National Socialism. Similarly, an exaggerated juxtaposition of the good with the bad can be misused to portray life and science under National Socialism simplistically as a series of clear choices between right and wrong, made by individuals who themselves fell clearly on one side or the other of the line between "Nazi" and "anti-Nazi."[221]

The historian Dieter Hoffmann has argued that if some of the scientists in the middle of the spectrum are critically examined—as this book intends—then there is a danger that they will be lumped together with the "real Nazis" and that the real differences between individual cases will be obscured.[222] In fact it must be possible both to criticize individuals standing somewhere between the two poles and nevertheless distinguish them from the more extreme examples at the spectrum's end. It must be possible to criticize or honor anyone according to objective criteria, no matter where they stand on the spectrum.

The political scientist Joseph Haberer characterized the behavior and self-image of scientists like Heisenberg as "resistance through collaboration."[223] In fact both sides of the struggle between "Aryan" and "Jewish" physics collaborated with the Third Reich. The former group supported the racist, anti-Semitic policies of National Socialism. The latter group helped the Third Reich wage its genocidal war. After the war both sides were convinced that they had thereby resisted the evil side of National Socialism.

If there ever was a "Nazi physicist," it was Johannes Stark. But despite his best efforts, in the end his science was not accepted, supported, or used by the Third Reich. In other words, his science was not "Nazi science." By the end of the Third Reich the followers of *Deutsche Physik* saw themselves as persecuted with any and all means.[224] Stark spent a great deal of his time during the Third Reich fighting with bureaucrats within the National Socialist state. Most of the National Socialist leadership either never supported Lenard and Stark or abandoned them in the course of the Third Reich.

Ironically Stark was just as concerned with science as with racism or political ideology. The race, nationality, or political standpoint of a physicist he attacked was at least in part a welcome excuse to be used to discredit a particular type of physics.[225] Stark's story also illustrates his stubbornness in pursuit of his goals. His science policy objectives in the Third Reich were practically the same ones he had had in the early twenties—except now combined with anti-Semitism and National Socialist rhetoric. The claims he made after the war of having fought against the excesses of National Socialism and for the freedom of research faithfully reflected his conviction that, during both the Weimar Republic and the Third Reich, he had done precisely that.

4

The Surrender of the Prussian Academy of Sciences

The three mutually exclusive categories described in the introduction, "Nazi," "anti-Nazi," or neither one nor the other, are equally problematic for scientific institutions. The Prussian Academy of Sciences (PAW), one of the first European academies of sciences, is one of the most notorious examples of a scientific institution going "Nazi." But a debate over whether or not the PAW should be labeled "Nazi" obscures both *how* and *why* it was transformed into a willing tool of National Socialism.

Most histories of the PAW under Hitler are dominated by three events from the early years of the Third Reich: Albert Einstein's well-publicized resignation, including the role played by Max Planck;[226] Max von Laue's successful efforts to keep Johannes Stark out of the academy;[227] and the takeover and transformation of the PAW by the National Socialist mathematician Theodor Vahlen.[228] Such portrayals of Vahlen are especially problematic because they can imply that the academy scientists were mere victims of an irresistible and ruthless perversion of their institution.

These three events are important, but when they are put into context, they reveal a more subtle picture. The institution and its members were both victims of and collaborators with National Socialism. This book will underscore both the victimization and collaboration of academy scientists with Hitler's movement by splitting the history of the academy during the Third Reich into two separate chapters. "The Surrender of the Prussian Academy of Sciences" examines the first years of the Third Reich, *before* Vahlen entered the academy. "A 'Nazi' in the Academy" begins with Vahlen's election and ends in the postwar era.

In contrast to what happened to the universities and the rest of the civil service, the transformation of the academy into a willing tool of National Socialist policy was a less gradual, steady loss of independence and scientific integrity. It was more of a blood-letting than a sudden wound. This transformation had two complementary components: the internal purge and restructuring of the academy according to the principles of National Socialism; and the external exploitation of the academy for National Socialist foreign policy.

The European academies of science date back to the seventeenth century. Some enjoyed the reputation of being the first scientific institutions. Leading scientists were honored through election to the academy and paid a salary, making them some of the first professional scientists. The academies published their transactions, including the research and achievements of their members, thereby becoming the first scientific journals. Academies corresponded and exchanged publications with each other, thereby facilitating international communication in science.

The PAW had two classes, scientific and humanistic, and three categories of members: ordinary or full, corresponding, and foreign corresponding. Only the full members had the right to vote on academy matters, but all of the members could present their own or someone else's scientific work to the academy. Even if this work did not appear in the academy transactions, the PAW thus provided an important scientific forum: the minutes of the academy meetings would establish scientific priority. Academies like the PAW also sponsored large-scale and long-term scientific projects, which were staffed by academy employees who were Prussian civil servants.

✠   ✠   ✠

The Einstein Affair When Adolf Hitler came to power in January 1933, Max Planck was perhaps the most respected and influential elder statesman for German science. His work on blackbody radiation and his quantum hypothesis—that energy exists in discrete, finite units, called quanta—earned him both a Nobel Prize and recognition as one of the founders of modern physics. Although his productive scientific work was now behind him, Planck dominated German science policy through a plethora of offices and responsibilities.

During the German Empire he received a professorship for theoretical physics at the University of Berlin, and was influential in the German Physical Society. In 1913, just before the start of war, he became rector of the University of Berlin. After the German defeat, Planck dominated the newly-founded Emergency Foundation for German Science by sitting on its committees and influencing how its money would be spent. In 1930 he became the president of the Kaiser Wilhelm Society. Most important for the history of the academy, in 1912 Planck was also elected one of the two standing secretaries of the PAW's scientific class. The four secretaries of the academy alternated every three months as its executive officer and acted collectively as its spokesman.[229]

Planck decisively influenced German science for decades despite radically different political and ideological regimes. He was in many ways a product of the Empire and saw service to the German state as subservient only to service to his science. During the heady early days of World War I, Planck allowed

The four Secretaries of the Prussian Academy of Sciences. Left to right: Heinrich Lüders, Ernst Heymann, Max Planck, and Max Rubner, ca. 1930. (From Ullstein Bilderdienst, Courtesy of the Library and Archives of the Max Planck Society.)

himself to be swept up in the enthusiastic and uncritical chauvinism shared by most German scientists. In contrast to many of his colleagues, Planck subsequently realized his error, and managed to accommodate both nationalism and scientific internationalism. As long as this war lasts, he said, Germans had only one task, serving the nation with all their strength. But there were domains of intellectual and moral life that transcended the struggles of nations. Honorable cooperation in science and personal respect for citizens of enemy states were compatible with "ardent love and energetic work" for one's own country.[230] During the Third Reich and especially during World War II, Planck would face the same dilemma: what to do when service for the German state came into conflict with service for the international community of science?

Although Planck personally rejected democracy, after German defeat in World War I he was willing to work with the Weimar

Max Planck speaking at the twenty-fifth anniversary of the Kaiser Wilhelm Society at the Harnack House in 1936. (Courtesy of the Library and Archives of the Max Planck Society.)

Republic for the good of his science. Thus it is no surprise that, at least at first, Planck was misled by the "national revolution" touched off by Adolf Hitler's appointment as German chancellor and the apparent return to traditional, authoritarian German values. Planck was always reserved towards the National Socialists and in time recognized that the new rulers were far more destructive toward science and society than the democrats had been. However, this realization was a difficult, gradual, and drawn-out process, arguably lasting until the very last years of the war when his son was murdered in the aftermath of the failed attempt to assassinate Hitler.[231]

Planck's role in the National Socialist transformation of the PAW is important because here is where he held onto his influence the longest. The struggle over the future of the academy was his last, most poignant stand against National Socialism. Unfortu-

Max Planck, date unknown. (From the E. Scott Barr Collection, Courtesy of the AIP Emilio Segrè Visual Archives.)

nately, Planck was already an old man in 1933, when he struggled to oppose his more vigorous and ruthless National Socialist opponents.

Planck and Einstein had a special personal and professional relationship. Despite Planck's political and ideological differences with the unconventional physicist, he respected Einstein's scientific talents so much that he arranged to bring him to Berlin before World War I. The package of appointments and benefits which successfully wooed Einstein included election as a full member of the PAW. The differences between the two physicists were exacerbated during World War I and the Weimar Republic, when Einstein's pacifism and subsequent support of the republic also made him the target of the far right in German politics.

Einstein was in the United States when the National Socialists came to power and immediately became a symbol for the Jewish "internationalist" influence which Hitler's movement was determined to eradicate. The political right, of which National Socialism was at first only a part, labeled anyone or anything internationalist which did not place the German nation first. Of course, Jews were by definition excluded from this nation. The reports of officially sanctioned anti-Semitism and the purge of the universities reached Einstein and appalled him. This led to his announcement that he would not return to Germany, which no longer enjoyed civil liberty, tolerance, and equality of citizens before the law.[232]

Planck, who typically was trying to work within the system in order to ameliorate the National Socialist policies and gain exemptions for a few Jewish colleagues, was not pleased by Einstein's action. In a private letter he chided Einstein: "by your efforts your racial and religious brethren will not get relief from their situation, which is already difficult enough, but rather they will be pressed the more."[233] Einstein's friend Max von Laue also criticized him privately for mixing science and politics: "but why do you have to take a *political* stand? I am the last person to criticize you because of your opinions. The political struggle requires different methods and different natures than scientific research. As a rule, the scholar is crushed under the wheels."[234]

Einstein replied by telling the PAW that, if he had defended Germany instead of criticizing it, then he would have contributed—if only indirectly—to the brutalization of morals and the destruction of all contemporary civilization.[235] Planck and the PAW urged him to resign, but Einstein had already done so. A day after his letter of resignation arrived at the academy, the Ministry of Education ordered the PAW to investigate whether Einstein had participated in the slander against Germany and if so to discipline him. The academy noted Einstein's resignation[236] and argued that any further action was moot.

But the politicized Einstein affair had gained too much notoriety and the Minister of Education, Bernhard Rust, was not content with a voluntary resignation. Rust insisted that academy secretary Ernst Heymann immediately take further steps. Heymann complied the very next day, issuing a public statement charging Einstein with slandering Germany and announcing that the PAW had no cause to regret Einstein's resignation.[237] The announcement came on the same day as the first quasi-official boycott of Jewish businesses in Germany.

The March 1933 elections, which were the last elections in the Third Reich and strengthened the National Socialists' position in the German government, were followed by attacks by the NSDAP rank-and-file, on individual Jews and Jewish businesses. This violence was not coordinated by the central government, rather was an example of the spontaneous "revolution from below" which terrorized the opponents of Hitler's movement and helped facilitate National Socialist efforts to consolidate their control over Germany.

Hitler sympathized with these attacks on Jews, but they threatened to get out of hand and jeopardize his alliance with Germany's conservative elites. Therefore Hitler decided to provide a controlled outlet for the energies of his rank-and-file and directed the party to organize a nation-wide boycott of Jewish business and professionals. Originally the boycott was intended to be indefinite, but concern about its negative impact on the economy and opposition by Reich President Hindenburg and the German Foreign Office persuaded Hitler to limit it to a single day.[238] The Einstein affair took place in this context: REM clearly wanted to demonstrate that it was doing its part in the struggle against the Jews.[239]

When the academy condemnation of Einstein was raised at a subsequent meeting, the academy retroactively approved Heymann's action and thanked him for his professional handling of the matter. However, Max von Laue did insist that the record show that no member of the scientific class had been consulted.[240] Planck did not dispute that Einstein had to go. Instead he regretted deeply that Einstein's political behavior had made his continuation in the

Academy impossible.[241] Planck apparently did not see, or did not want to see, that eventually Einstein and all Jews would be forced out of the PAW.

Although Planck went along with the censure of Einstein, he also did what he could to soften posterity's judgment by lauding Einstein's work before the academy as comparable only with Kepler's or Newton's.[242] In response to Planck, Heymann replied that he had been aware both of Einstein's great scientific significance and the consequences his expulsion would have. For this reason he had consulted men with foreign policy experience.[243] In fact, von Laue and Planck were most concerned with separating science and politics. When confronted by the National Socialist purge of Einstein and the academy's acquiescence, they insisted that there was no scientific or professional justification for it.

Planck and other academy officials may well have been reluctant, non-enthusiastic participants in the Einstein affair, acting for what they considered prudent and pressing reasons.[244] But despite the fact that a few individuals like von Laue took Einstein's side, and others like Planck regretted the incident, the majority went along with the wishes of their government.[245] Whatever Planck's motives might have been, the public effect of the Einstein affair was clear. Within Germany, the PAW shared in the official ostracism of Einstein; outside of Germany, the PAW was a willing accomplice of National Socialist anti-Semitism.

                    ✠   ✠   ✠

Barring the Door to Johannes Stark Max von Laue did his Ph.D. with Planck and his _Habilitation_ with Arnold Sommerfeld in Munich, where he discovered x-ray interference in crystals and thereby earned the 1914 Nobel Prize.[246] In 1909 von Laue was so eager to return to Berlin and rejoin Planck that he traded his full professorship in Frankfurt for an associate professorship[247] in the Reich capital. Von Laue had actively and publicly defended Einstein and his science when they were attacked in the early twenties, and continued to oppose _Deutsche Physik_ in the Third Reich.[248]

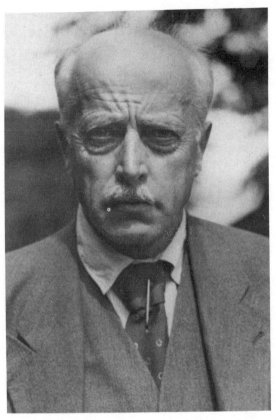

Max von Laue, 1945 at Farm Hall. (From the National Archives and Records Services.)

In November 1933 Johannes Stark, the Nobel laureate and enthusiastic National Socialist, was proposed for membership in the PAW. This was a distinction which Stark normally would have a right to expect, thanks to his recent appointment as president of the Imperial Physical-Technical Institute.[249] Government officials pressured the physicist Friedrich Paschen to nominate Stark and the academy to elect him.[250] But Stark's old adversary, von Laue, openly opposed Stark's admission, despite the latter's obvious political influence.[251]

Von Laue told the academy that in the past he had watched with regret as Stark was passed over for appointments, including to the academy, even if it was partly his own fault. But Stark had recently called for a dictatorship of physics and threatened to use force against anyone who resisted him. The academy tabled the proposal and thereby excluded Stark. He responded in December 1933 by canceling von Laue's position as a scientific advisor to the PTR.[252] Stark had made plans to fire von Laue before the academy affair was decided, but the timing now seemed especially appropriate.[253]

Von Laue's opposition was courageous and principled, but why was he successful? In contrast, von Laue did not oppose Theodor Vahlen and Eugen Fischer when they were proposed for academy membership in 1937.[254] Fischer was a race hygienist and respected anthropologist who had placed his expertise in the service of the National Socialist state. The mathematician Vahlen, a National Socialist of even longer standing than Stark, was also both anti-Semitic and anti-Einstein.[255]

Stark's personality may have been harder to swallow than Vahlen's, but the latter obviously represented an equal if not greater threat to German science. However, in striking contrast to Stark, Vahlen's high-ranking position in REM and his membership in the SS gave him real political power. Germany in 1937 was also very different from 1933. Gestures of opposition which could be made in the first year of the Third Reich were much harder even to contemplate four years later. Von Laue and others could oppose Stark without grave repercussions, even though political allies had pushed his candidacy. But the academy had to submit to Vahlen.

Historians often emphasize von Laue's opposition to National Socialism. For example, Alan Beyerchen argues that this physicist rejected even a show of cooperation.[256] Unfortunately, even von Laue had to make concessions to National Socialism. He certainly did not resist the National Socialists in general as vigorously as he did Stark, and even his courageous opposition to Stark was possible only because the latter had so many enemies within the state bureaucracy. When National Socialist officials assessed the mathematicians and physicists at the University of Berlin at the end of December 1934—thus after the academy had rejected Stark—they judged that von Laue was an excellent scientist. Pedagogically he was less talented, and nothing was known about his political conduct.[257]

Furthermore, von Laue sometimes had to make concessions to Stark. Von Laue was in charge of the Physics Colloquium at the University of Berlin, where in the past members of the PTR had been valuable participants. This cooperation was now threatened by the open hostility between von Laue and Stark. Von Laue decided to cooperate and compromise with Stark. It would bring von Laue great pleasure, he wrote, if Stark would give his blessing and thereby support to this type of cooperation between the PTR and the university. Moreover, von Laue signed the letter "Heil Hitler!"[258] The point here is not to accuse von Laue of being a "Nazi," rather to illustrate how difficult it was for anyone or any scientist to avoid some sort of submission to or collaboration with National Socialism.

⊞   ⊞   ⊞

The Purge The National Socialist transformation of the PAW included four complementary strategies: purging the academy of racial and political opponents; coercing the real or apparent allegiance of the remaining members; bringing scientists into the academy who actively supported National Socialism; and, perhaps most important, allowing a great deal of business as usual and thereby encouraging members to believe or hope that government intervention was over or would soon end, leaving them in peace.

On 7 April 1933 the German government announced the infamous "Law for the Restoration of the Professional Civil Service," the legal framework for the purge of the German government of all racial and political enemies or opponents of Hitler's regime.[259] The euphemistic title of this legislation implied that the

Weimar Republic had debased the bureaucracy since World War
I and cynically portrayed this purge as a restoration.

This law had an important effect on German science because
all university teachers and most other researchers were civil servants. The National Socialists adopted this tactic because of its
apparent legality. By providing a law for their purge of the civil
service, the new government won the support of many Germans
who otherwise might have protested or at least condemned the
dismissals. So long as the National Socialists could cloak their
racist and ideological politics in legality, they could count on the
passive acceptance and tacit support of many Germans who themselves were not racist, but who were nevertheless unwilling to
question their government.

Civil servants who had been hired after 1918, who were
"non-Aryans"—which at this time meant having at least one Jewish grandparent—or because of their previous political activity did
not ensure that they would act at all times and without reservation
in the interests of the national state, could be fired or retired—and
usually were. If none of these categories fit, an official could still
be dismissed by means of the cynical justification of rationalizing
the administration. On 30 June, this so-called "Aryan paragraph"
was extended to officials married to "non-Aryans."

The April 1933 civil service law included some exceptions for
"non-Aryan" civil servants who had fought in World War I, but
even these were not always honored and were all eventually
rescinded. Although this policy caused a great deal of personal
hardship, the overall quantitative effects of this purge were com-
paratively small, which illustrates how homogeneous, "Aryan,"
and conservative the civil service had been.[260]

The National Socialists were very thorough and took great
pains to ensure that no one could fall between the cracks. REM
decreed that no official under its jurisdiction could be given a leave
of absence and sent abroad without its permission. The ministry
thereby eliminated one way officials tried to avoid firing someone,
especially if they believed that these excesses would soon blow
over. However, REM also noted that the Aryan paragraph should

not be applied to areas for which it was not intended, in particular, not to the private economy.

This policy reveals the limits of the National Socialists' power at this time. The civil service could be purged, but the private sector had to be left alone. Thus even the well-publicized boycott of Jewish businesses on 1 April 1933 was called off by the party leadership after only one day and not repeated. Only in the aftermath to the infamous "Night of Broken Glass" in 1938 did the National Socialist state take further steps to force out Jewish businesses.[261]

The purge of German bureaucracy did not stop with the civil service law. In June REM passed on a decree from the Reich Ministry of Interior that any civil servants who were Jehovah's witnesses were to be fired because they could not be counted upon to serve the national state unconditionally and at any time. This decree was one of many confidential, i.e., secret instructions which were to be carried out, but not made public.[262] At almost the same time, the Interior Ministry instructed all state institutions that, until further notice, the following civil servants should not be promoted: individuals who had belonged to the Social Democratic or liberal political parties; who had opposed the national renewal; who were not pure "Aryans"; and who were married to "non-Aryan" women.[263] Thus even if an official had not been fired, he might be denied all hope of further promotion or advancement.

Many former civil servants still believed in the German legal system and went to court in the hope of reinstatement. In July 1934, the Prussian Ministry of Justice made clear that the many legal cases brought by civil servants fired or forced into early retirement by the Civil Service Law would be dismissed. Moreover, state institutions like PAW were to send any information they had on such matters directly to the Minister of Justice.[264] Individuals who persisted or made trouble not only would not get their jobs back, they could face even worse treatment.

If the politically motivated purge was not enough, in the autumn of 1934 REM announced cost-cutting measures which also directly affected the PAW. First, civil servants who were not fulfilling a necessary function would be fired. Second, almost everyone would receive a cut in pay. Although the academy members had not officially protested the Civil Service Law, they now instructed their secretaries to complain about the cuts to the Ministry of Finance, but with little hope of success.

The National Socialists took additional steps in 1935 to tighten their hold on the bureaucracy. Each civil servant had to provide written documentation of his "Aryan" ancestry.[265] In the fall REM ordered all civil servants to submit a written list of all professional organizations they had belonged to or were still members of since the end of World War I.[266] Since the list of politically suspect organizations increased over time, such information inevitably led to more resignations and dismissals. An October 1935 REM decree directed the academy and all other agencies under its authority to suspend immediately all remaining civil servants who were Jewish, or had three or four Jewish grandparents.[267] In December another order added insult to injury by decreeing that if civil servants who were politically suspect or "non-Aryans" resigned or even retired after twenty-five years of service, then they could not be thanked officially by ceremony or letter.[268]

The National Socialists also took care that any new appointment fit their specific requirements. In December 1935 PAW was informed that when a candidate for the civil service was proposed to REM, the proposal had to include the following information: ideological conduct and conviction, efforts on behalf of National Socialism and in what form this took place, attitude towards duty, professional abilities, camaraderie, and other positive and negative character and professional qualities.[269]

As far as the existing bureaucrats were concerned, civil servants could only be promoted (and thereby given a raise in pay) if their past political stance and their conduct since 1933 ensured that they would at any time and in every way fight for and effectively represent the National Socialist state.[270] Thus in a step-by-step fashion the National Socialists molded a compliant and subservient civil service.

In June 1933 REM decreed that the civil service law would be applied to the employees of the PAW. Although most of the full members were subject to the law in their capacity of university professor, for the moment the position of "non-Aryan" academy members was not threatened. When Planck and the other three academy secretaries met to discuss this matter, they were pleased to note that none of the civil servants or other paid employees were affected. The academy had only "Aryan" employees. However, when a subsequent decree extended the law to unpaid employees as well, one individual was affected. The academy sent him the official questionnaire and washed their hands of him. He had to make his own case to REM for remaining at his post.[271]

The German academies in Berlin, Göttingen, Heidelberg, Leipzig, and Vienna had formed an academy cartel during the Weimar Republic as a response to what they considered the international boycott of German science.[272] When the cartel met in June 1933, the political upheaval with its still unpredictable effects lay heavy on their minds. Fortunately for the academies, they had a record of consistently and decisively taking a nationalistic stance in their struggle against the foreign policy of the Weimar Republic. The Vienna academy in neighboring Austria hastened to declare its loyalty to and solidarity with the Reich members of the cartel.

The academies were faced with the now acute "Aryan question." The universities and the rest of the civil service were being ruthlessly purged of scientists and scholars either racially or politically objectionable to National Socialism. There was no reason to expect that the academies of science would fare any differently. Although the government had not yet taken any step, the questionnaires would certainly come. The Austrian representative remarked that, although the National Socialist policy did not affect them in Austria, in the future the Austrians would be much more demanding and cautious with regard to the election of "non-Aryan" members. Finally, PAW Secretary Heinrich Lüders brought up a matter of great concern: English newspapers had exhorted the foreign corresponding members of the German academies to resign in protest. Fortunately for the German academies, these members had not yet done so.[273]

This meeting was the first of many discussions of a crucial dilemma for the German academies in general and the PAW in particular. The academies were completely dependent on government support. Cooperation with the National Socialists meant making concessions on the "Aryan question." But such measures also threatened to provoke mass resignations of their foreign corresponding members, which in turn fundamentally threatened the academies themselves. If the academies became showcases of "Aryan" science in Germany, then they would no longer be accepted by the international scientific community.

Shortly before Christmas 1935, news reached the PAW of unrest in their sister academy in Heidelberg, by far the most radically National Socialist academy. Three younger members in Heidelberg announced their intention of giving talks before the academy on the new type of scientific research, i.e., race-based science, but added that the presence of the "non-Aryan" members would be embarrassing and hinder their appearance. Thus they proposed that the Jewish members either resign or agree not to attend academy meetings in the future. The embattled "non-Aryans" refused to leave unless the entire academy asked them to go. The young radicals then backed down, at least temporarily, and the Heidelberg academy passed the matter onto the cartel, which delayed making any decision as long as possible.[274]

✠   ✠   ✠

Affirmative Action for National Socialists Once the National Socialists had purged the bureaucracy of their obvious enemies, they began using civil service jobs as rewards for their long-standing supporters. In August 1935 REM ordered PAW to report how many of its employees had joined the NSDAP (National Socialist party). Unfortunately, the academy had none to report.[275] In January 1936, REM specified who was to be favored: only applicants who had joined the party before 14 September

1930—well before Hitler's movement appeared heading for power—were to be given preferential treatment.

However, at almost the same time REM ordered the PAW and all other agencies under its control to make an annual report on the following: employed NSDAP members who had joined the party before Hitler's appointment as German Chancellor; [276] the number of National Socialists hired as civil servants; the number of applications from such individuals turned down because of lack of positions; and the number of these National Socialists who had been unemployed. Even if no such individuals had been hired or had applied, the PAW nevertheless had to submit a written report. Indeed the PAW had to report once again that no such individuals were employed.[277] These were only the first of the many regular inquiries which pressured the PAW to employ NSDAP members and coerced existing employees to join.[278] Eventually many academy employees and a significant minority of full members joined the party.

✠   ✠   ✠

Coercing Allegiance There were other, more subtle ways to transform the PAW. In February 1934, REM ordered the PAW to close all official correspondence with the words "Heil Hitler!"[279] In September 1935 the Heil Hitler! formula was extended to special celebrations and congratulations, although it was not to be used in correspondence between state offices.[280] This technique forced conformity. Either someone refused to use Heil Hitler! and thereby revealed himself as an enemy to be dismissed or he went along with the mandatory formula, and apparently supported the regime and the Hitler cult.

Hitler's power was limited for the first year and a half of the Third Reich by the presence of Reich President Hindenburg. There was a danger that the Army, the only part of the German state which could still topple Hitler in a coup, would insist upon replacing the aging president and thereby thwart Hitler's ambitions of achieving total power. Hitler bought the support of the Army in late June 1934 with his bloody purge of the National Socialist SA, a potential rival of the traditional armed forces.[281]

When the President and former war hero finally died in August, Hitler fused the offices of Chancellor and President into his new title: *Führer*, or "leader." The National Socialists then honored the deceased Hindenburg as part of their strategy to minimize the opposition to Hitler's consolidation of power. Interior Minister Wilhelm Frick and Propaganda Minister Josef Goebbels subsequently decreed that all civil servants participate in the two-week period of mourning for Hindenburg by wearing a mourning flower on the left arm.[282]

Three weeks later, the Reich government took additional steps to bind its civil servants to it by means of a new oath for all governmental employees:

> I swear that I will be loyal and obedient to the *Führer* of the German Reich and people, Adolf Hitler, respect the laws, and exercise the obligations of my office conscientiously, so help me God.[283]

This oath provides another example of the elaborate, totalitarian mechanisms the National Socialists used in order to ensure compliance. All civil servants had to swear this oath. Each institution was required to send the Ministry of Interior a written report on the oath-taking within eight days.

Moreover, the oath had to be taken in a certain form. The officials and employees gathered together, the head of the institution read aloud the oath, and the civil servants repeated the oath in unison. Each civil servant immediately confirmed his oath in writing, a copy of which would remain in his personnel file. Any civil servants on leave had to take the oath immediately upon their return. REM ordered that all official trips or other reasons for a civil servant's absence be postponed until after the swearing-in ceremony.[284]

Thus elaborate steps were taken in order to ensure that everyone take the oath of allegiance to Hitler. It was an integral part of the duties of a civil servant; refusal to take it was grounds for forced retirement or dismissal. If a civil servant took it with any reservations, then that would be equivalent to refusing the oath. Finally, the ministry passed on a thinly veiled threat. Anyone who was not prepared to follow his oath without reservation should resign. If he did not, then he should expect to be treated the same as those who had flatly refused to take the oath.[285]

The National Socialist state was concerned that all Germans take part in public National Socialist rituals as part of the "peoples' community"[286] and thereby at least appear to express solidarity with the Third Reich. In November 1934 REM decreed that appointments and promotions of all employees would be announced on one of the new national, i.e., National Socialist, holidays. April 20, Hitler's birthday, was especially suitable.[287] The PAW was also caught up in this ritual tribute to National Socialism. On 30 January 1936, the anniversary of Hitler's appointment as Reich Chancellor, rotating chairman Planck began the day's business by reminding the members of the national significance of the day.[288]

Such concessions to the regime had a similar manipulative and exploitative effect as the Heil Hitler! salute. Thus Planck and his colleagues were coerced into public gestures of support for National Socialism. The 1936 NSDAP party rally in Nuremberg throws some light on how daily life in the PAW had been changed by the new order. The entire academy was ordered to gather at 4:25 PM on 28 September for a communal broadcast of a Hitler speech which would begin five minutes later. Furthermore, each academy employee or voluntary co-worker had to sign the memo informing them of the communal action, thereby eliminating any excuse for not attending.[289] Such communal meetings were common in the Third Reich, and were yet another technique to coerce conformity. If someone did not participate, or attended and protested, then he would reveal himself as an enemy of the regime. If he did participate, then he gave the appearance of solidarity with Hitler's movement and thereby helped justify it.

✠   ✠   ✠

Business as Usual However, daily life for the members of the PAW—as opposed to its employees—during the first years of the Third Reich probably appeared quite normal and apolitical. For example, there obviously was no censorship of Einstein's science. On 10 January 1935, von Laue presented a scientific paper by a colleague which applied Einstein's theories to cosmology.[290] In April 1936, von Laue delivered a paper on the quantum theory— yet another branch of physics that had been labeled "Jewish science."[291] These topics were at the cutting edge of science, but such gestures by von Laue may also have been intended to make up for acquiescence with regard to other matters. Yet a month later von Laue drafted a congratulatory letter from the PAW to Philipp Lenard on the fiftieth anniversary of his doctorate.[292] Von Laue was either making concessions to, or studiously ignoring the political nature of *Deutsche Physik*.

✠   ✠   ✠

The National Socialist Fifth Column The PAW was not merely attacked and pressured from the outside, it was also betrayed to the National Socialists from within. Perhaps the first indication of a National Socialist fifth column within the academy came when the respected mathematician Ludwig Bieberbach proposed in early 1935 that in the future the PAW ask REM's permission before electing corresponding members from foreign countries.[293] The academy secretaries rejected this suggestion,[294] which in effect would have surrendered the academy's autonomy.

Bieberbach was no old fighter, rather a classic example of an opportunist who embraced National Socialism once it came to power. He had held the position of full professor of mathematics at the University of Berlin since 1921 and full academy membership since 1924. Bieberbach's colleagues and students were surprised when he turned to the National Socialists in 1933; he had given no indication during the Weimar Republic of fascist sympathies. He joined the National Socialist University Teachers League in November 1933, the NSDAP in May 1937, and belonged to

Ludwig Bieberbach, date unknown. (Origin unknown. Published in Herbert Mehrtens, "The 'Gleichschaltung' of Mathematical Societies in Nazi Germany," *Mathematical Intelligencer*, 11, No. 3 (1989), 48–60.)

several other National Socialist organizations, including the SA (Stormtroopers).[295] After the start of the Third Reich, Bieberbach was rewarded for his political cooperation with the appointment as dean of the scientific faculty at the University of Berlin.

Bieberbach made his reputation as the "Nazi" among mathematicians by the theories on the psychological (and thus racial) background of different mathematical styles which he propagated after 1933. According to Bieberbach's *Deutsche Mathematik* (literally translated as "German Mathematics"), "Aryans" and Jews created different types of mathematics because they belonged to different races. Thus he advocated a philosophy of science analogous to the *Deutsche Physik* of Lenard and Stark, even though he did not support the two physicists in the politics of the Third Reich.

Bieberbach also tried and failed to seize control of the German mathematics community by taking over its professional organizations. His mathematician colleagues managed to thwart Bieberbach's ambitions by making concessions to other National Socialists. By 1937 Bieberbach and his group were an "ideological residue" in the system of mathematics without substantial influence.[296] They continued to publicize their *Deutsche Mathematik* as an example of true National Socialist science, but just like the *Deutsche Physik* of Lenard and Stark, Bieberbach's group was ignored by the National Socialists bureaucrats in charge of science policy. But if Bieberbach had failed to realize his aspirations for German mathematics, he could still work to transform the PAW along National Socialist principles.

On 30 September, 1935, REM followed Bieberbach's suggestion and decreed that German scientific organizations had to proceed very cautiously when naming foreign scholars as corresponding members. The academy now had to take care that only scholars who would at least take a neutral stance toward the new Germany be considered. If a case was questionable, then the academy should contact the ministry ahead of time. The PAW responded that it had always taken the political stance of the potential candidate into account when electing corresponding members and would certainly do so in the future.[297]

In early 1936, three academy members who supported National Socialism, Bieberbach, Hans Ludendorff,[298] and Paul Guthnick, proposed that the PAW fill two free positions with representatives of anthropology and racial science, scientific disciplines which were especially important to National Socialist ideology. They suggested Eugen Fischer and Hans F. K. Günther. The former was a respected anthropologist and leading race hygienist ("race hygiene" was the German term for eugenics); the latter was a popular racial theorist who invented a typology of racial types which facilitated and justified racist policies as well as Bieberbach's *Deutsche Mathematik*.

At first the academy responded favorably,[299] but a few weeks later the secretaries announced that it would be better not to constrain the academy by binding positions to particular scientific disciplines.[300] This response was probably an attempt by the academy to retain some of its steadily eroding independence and scientific standards for membership. It was willing to elect such scholars, and indeed did subsequently bring Fischer into the academy, but also wanted to avoid sanctioning particular types of science.

At this time the academy still enjoyed its independence with regard to the election of members, although some concessions had been made. In 1935 Karl Becker, an Army officer interested in science policy, was elected a full member of the PAW.[301] The Reich Minister of War subsequently thanked both REM and the PAW.[302] The election of a member of Germany's conservative military elite was no doubt welcomed by some as insurance against an invasion of the PAW by radical National Socialist elements, but it nevertheless represented a profound break with tradition. Someone like Becker probably would not have been elected as a full member during the Weimar Republic or even the militaristic German Empire.

In late April 1937 PAW members proposed the IG Farben industrialist Carl Bosch for honorary membership in the academy. When Bieberbach raised objections, perhaps because Bosch was ambivalent about National Socialist policies, the vote was postponed.[303] However, Bosch was about to be elected President of the Kaiser Wilhelm Society with the approval of Minister Rust and obviously had his backers in the National Socialist state. When the vote was finally taken, Bosch received only one opposing vote.[304] Members like Becker or Bosch rarely if ever participated in the academy. Their elections, like the subsequent elections of leading

National Socialists, were merely insincere and increasingly mean-ingless honors designed to curry political favor and contributed to the scientific debasement of the PAW.

In early 1936, without warning REM began to restructure the Reich academies along National Socialist lines, simultaneously forbidding any public discussion of this reform. The Bavarian Academy sent PAW a copy of the official publication of laws for the state of Bavaria, which included significant changes in the organization of the Munich academy. The president of the acad-emy and the two secretaries of each class, who had previously been elected by the academy for three years, would now be appointed by Reich Minister Rust. The Munich academy had received no notice of these changes other than the publication of the law itself,[305] a common and effective strategy employed by the Na-tional Socialists to create confusion and minimize opposition to their policies. The members of the Berlin academy must have asked themselves whether they would be next. A few weeks later REM strongly suggested that PAW should expect a similar reorganiza-tion.[306]

The academy took the hint and chose to do voluntarily what they assumed would otherwise be done by force. On 27 February, the four secretaries reported to the full academy their proposal for altering the academy statutes. The most important change was a simple one: the word "elected" was replaced by "appointed." Although the PAW would continue to nominate and elect scholars and scientists as before, and thereby preserve the illusion of inde-pendence, in fact the results of their elections now became mere recommendations which officials in REM could either accept or reject. The academy had in effect surrendered their independence, and indeed went far beyond the changes forced upon the Bavarian academy. After a short debate on the secretaries' action, the acad-emy approved the proposal unanimously.[307]

The threat of a public debate over Jews in the academy was one reason why the German academies were willing to give up some of their independence before it was required. The Einstein affair was still fresh on everyone's mind, and since the PAW still had Jewish members, the academy was vulnerable. When the PAW was attacked in the January 1937 issue of the National Socialist journal *Volk im Werden* for harboring Jews and opposing National Socialism, the academy empowered its secretariat to investigate the matter and bring it to the attention of the ministry.[308] A month later an academy member pressed the matter further and insisted that they could not remain silent about this attack.[309]

But REM decided that it would be counterproductive for the PAW to clash publicly with their critic, the Heidelberg anthropology professor Ernst Krieck, who had gone so far as to question the academy's right to exist.[310] When REM did respond officially, it hardly calmed the academy. Rust chastised Krieck for going outside of official channels, but welcomed any suggestions he might have for renewing and reorganizing the PAW. This was a question that had been occupying Rust himself for a long time.[311]

In February 1937, PAW was finally confronted with something it must have seen coming: the forced expulsion of its Jewish members. All the German academies were now directed to report how many "non-Aryan" members they had, when these members had been elected, which of the honorary and corresponding members were Jewish, and what steps could be taken against these latter individuals.[312] Thus from the very beginning REM pursued a policy that ensured that the PAW and its members would be accomplices to any purge.

REM appreciated the fundamental problem of the "non-Aryan" foreign corresponding members. All concerned wanted to retain at least the appearance of legality, but legally these members could not be dismissed for being Jewish unless they were sent questionnaires and required to prove their "Aryan" ancestry. Any such action would most probably lead to a mass exodus of foreign corresponding members from the German academies, generate a great deal of bad publicity, and make the PAW less valuable to the National Socialist state. Thus the responsible REM official even went so far as to forbid the PAW to take any such measures on its own without explicit authorization.

However, the same bureaucrat asked a PAW member whether it was true, as had been reported, that Einstein was still a corresponding member? The PAW representative hastened to describe the events of Einstein's dismissal and assure the official that Einstein no longer had anything to do with the academy. REM wanted to handle the purge of "non-Aryan" academy members quietly, perhaps by dissolving and reconstituting the academy, thereby reconfirming all existing members while omitting the Jews. This common bureaucratic tactic during the Third Reich would allow the government to obscure its brutal personnel policy.[313]

Although the academy had not been previously included in the purge of Jews, most of its members were also active university professors or other types of civil servants who had already been required to demonstrate their "Aryan" ancestry in order to retain their jobs. The "non-Aryans" who were fired usually left Germany and thereby the PAW. But there were still a few older scientists left in the academy. On 1 March 1937, PAW sent its report on "non-Aryan" members to REM. There were three "non-Aryans" among its sixty-three full members. A fourth member was one-quarter Jewish. All the corresponding members in Germany were university professors or state civil servants who had already demonstrated their "Aryan" ancestry.

The PAW was in no position to provide or determine the required information with regard to their foreign honorary and corresponding members. There were five exceptions, and they were corresponding members who had previously been dismissed from their university positions as "non-Aryans." The draft report closed with a passage that was crossed out and omitted from the final version, but illuminated the PAW's attitude towards Jews; these figures also showed how reserved the academy had always been toward admitting "non-Aryans."[314]

In April an exceptional meeting of the academy cartel was held to discuss their "non-Aryan" members. All agreed that merely asking whether or not a foreign corresponding member was "Aryan" would probably lead to a mass resignation. However, they also recognized that the dismissal of the remaining "non-Aryan" academy members, which appeared more and more likely, would probably lead to the same thing. All agreed that the loss of their foreign members would make impossible the traditional function of the academies—the cultivation of scientific relations between Germany and foreign countries.

Finally, the formal cartel response emphasized how small the number of "non-Aryan" members was and how little justified the accusation that the academy had been "Jewified." The few remaining "non-Aryan" members, who were insignificant and hardly noticeable to the public, were much less dangerous for Germany than the consequences of expelling these members. The cartel academies considered themselves responsible for Germany's foreign scientific relations. They had the duty to warn of potential damage to these relations, and also had the right to be heard when decisions were being made which would prevent the academies from fulfilling their unique and most important function in the life of the nation.[315]

The members of the PAW approved sending the cartel report on to REM with one dissenting vote from Bieberbach.[316] The unity of the cartel also fell apart on this point: the more radical Heidelberg academy refused to go along and submitted its own report.[317] When secretary Heinrich von Ficker subsequently discussed the matter of the remaining "non-Aryan" full and corresponding members with the responsible ministry official, he found the latter very understanding, but also saw clearly how difficult this matter was. Domestic and foreign policy considerations were often very difficult to reconcile.[318] It was clear that sooner or later the Jewish members would have to go.

The Prussian Academy of Sciences was not seized or taken over by the National Socialist mathematician Theodor Vahlen or the Third Reich. When faced with a choice between endangering their academy or acquiescing in the racist purge of the PAW, the academy scientists surrendered their independence and became accomplices by helping the National Socialist state force the Jewish scientists out of the academy. No "Aryan" scientists resigned in protest. Indeed there is no record of a scientist even considering resignation. The academy report on the "Aryan question" did argue against the purge, but only because it would make the work of the PAW more difficult, if not impossible. No one was willing to question publicly the fundamental National Socialist principle that only "Aryan" scientists deserved to be in an academy of science.

5

A "Nazi" in the Academy

The "Little Hitler" in the Academy In February 1937 the scientific class nominated the mathematician Theodor Vahlen and the race hygienist Eugen Fischer for election to the academy.[319] Bieberbach and Planck were among the sponsors of both proposals.[320] Although Fischer's science, anthropology, and eugenics, were more relevant to National Socialist science policy, Vahlen had extremely impressive political credentials for the Third Reich, even better than Philipp Lenard or Johannes Stark. Vahlen was born in 1869, was a decorated veteran of World War I, and had been a member of the NSDAP from the very beginning. He served as regional leader for Pomerania and member of parliament during the twenties, joined the Stormtroopers in 1933, and switched over to the SS in 1936.

Vahlen became full professor of mathematics at the University of Greifswald before World War I and university rector in 1924. Moreover, Vahlen was one of the few professors in the Weimar Republic to embrace early and openly Hitler's movement. In 1924 Vahlen incited a crowd at a rally against the republic and took

Theodor Vahlen, 1934. (Courtesy of the Ullstein Bilderdienst.)

down the Prussian and Reich flags from the University flagpoles. The republican government immediately placed Vahlen on leave and eventually fired him without a pension for political abuse of his function. Vahlen was offered a professorship outside of Germany, at the Technical University in his birthplace, Vienna.[321]

Vahlen was also a respected, although not first-class, mathematician. His main interests lay in the areas of ballistics and nautical navigation. During World War I he had led an artillery battery. Devastating criticism in 1905 from a Jewish colleague not only pushed Vahlen into applied mathematics, what he characterized as the natural, concrete way of thinking of the "Aryan" race, but may have made him more anti-Semitic. As early as 1923 Vahlen characterized mathematics as a mirror of the races.[322] In 1934 Vahlen began his close collaboration with Ludwig Bieberbach to propagate *Deutsche Mathematik* through a journal of the same name.

But Vahlen also tried to use more rational arguments in the service of National Socialist science policy. For example, Vahlen was more circumspect than the adherents of *Deutsche Physik* on the subject of the theory of relativity and took care to use scientific arguments when attacking Einstein's work. In 1933 he responded to a proposal that this theory and its supporters be forcibly eradicated by insisting that to use the Education Ministry's power in this matter would mean regressing back to medieval methods. The National Socialists would be more successful in the purification and clarification of their spiritual life by placing the best men in the best positions.[323] Eventually Vahlen adopted the common tactic of ascribing the theory of relativity to other "Aryan" physicists, thereby accusing Einstein of plagiarism, but also making the theory palatable to the National Socialist state.[324]

Vahlen gained power and influence over science policy in the Third Reich mainly because he was a fascist, not because of his mathematical prowess. In March 1933 Vahlen was appointed to the University Division in REM. A little more than a year later he was in charge. He was especially active in implementing the Law for the Restoration of the Career Civil Service and decisively molded the Ministry's science policy towards the Kaiser Wilhelm Society, the PAW, and the Research Foundation. On 1 January 1937 Vahlen was relieved of his duties in the Ministry.[325] As his subsequent conduct would show, Vahlen was most probably eased out because he was no longer able to fulfill his function.

In the spring of 1936 Vahlen tried to take over the Kaiser Wilhelm Society through the back door. The mathematician sent an emissary to Philipp Lenard and asked him to accept the presidency of the Society as a figurehead. Vahlen would do all the work.[326] Lenard replied that Vahlen should take the job himself.[327]

If he could have, he probably would have, but the Society had influential allies within the National Socialist state. Lenard could arguably have been pushed through in the face of opposition, but not Vahlen. One of Vahlen's successors at REM hinted to Johannes Stark that Vahlen had been forced to give up his position in the ministry.[328] In any case, Stark believed that Vahlen, who in his opinion had little understanding or character, wanted to become president, not to further a National Socialist revolution in science, but instead out of desire for money.[329]

It was no coincidence that Vahlen was nominated for the PAW after his efforts to manipulate the presidency of the Kaiser Wilhelm Society had been thwarted. Vahlen's entry into the academy was coerced by his National Socialist allies. As usual, two competent experts assessed his scientific career and justified his admission, but made clear in subsequent publications that they in fact thought little of the very work they had previously praised.[330]

However, Vahlen's election was complicated by the traditional method of voting in the PAW. New members had to be nominated within a class, elected by that class, and finally elected by the academy at large. All these votes were taken by a special form of secret ballot: each member would place either a black or white sphere in a container. If the candidate received a large enough majority of white spheres, then he was elected. The spheres posed problems for scientists who were intent on transforming the PAW into a National Socialist institution, but at the same time wanted to keep the appearance that the long-standing traditions of the academy were still being respected.

When the vote on Fischer's and Vahlen's candidacy was held on 15 April , it ended with a shocking result. Although Fischer was elected by a wide margin, Vahlen did not achieve the necessary majority.[331] Such a defeat was almost unheard of at the PAW, and revealed how problematic the black and white spheres could be: there was no way to stop a member from professing support in public but casting the black sphere in secret. For example, although Planck had been one of the sponsors of Vahlen's appointment, he could nevertheless have secretly voted against him.

However, Vahlen and his supporters were not finished and the victory of Vahlen's opponents proved short-lived, if not counterproductive. Bieberbach immediately called for the following changes in how the PAW elected members: only the members of the relevant class would vote; each member would be asked for his opinion publicly, i.e., no more secret ballots; and the secretary alone would then decide whether or not this name should be proposed to the ministry.[332] Less than a month after he made this threat, Bieberbach simply started the process all over again. Vahlen was proposed by several members of the scientific class,[333] nominated by a wide margin,[334] and on 24 June was finally elected by a sufficient majority.[335]

It was also no coincidence that Vahlen retired from REM a few months later, when he received the unusual honor of a personal letter of congratulations from Hitler.[336] Vahlen was still an honorable long-standing National Socialist activist, but he had gotten older and had noticeably slowed down. When the SS accepted him in 1936, the SS Security Service pointedly requested that he not be assigned to them.[337] As far as the SS and REM were concerned, the academy was a suitable rest home for an aging old fighter. In contrast to the more powerful and independent Kaiser Wilhelm Society, the PAW could not resist a takeover.

In October 1938 Minister Rust informed PAW that the statutes of the academy would be changed corresponding to the fundamental ideology of National Socialism. The leadership principle had to be introduced, thereby installing a strict hierarchy and eliminating any remaining democratic elements. The structure of the academy leadership would be altered to include a president, vice president, and two secretaries, one for each class. One of the two secretaries would also handle the business of the entire academy and have the title of General Secretary.

The number of full members would be expanded, which was an effective way to create a majority of National Socialist members while retaining a sense of continuity with the old academy. REM not only had to approve the election of all members, but the PAW had to report its nominees to REM before any public announcement was made. Election to the academy was also no longer permanent. REM could withdraw its approval of a given member at any time.

Full members could be only Reich citizens, i.e., "Aryans," who lived in Prussia.[338] The Reich Citizenship Law had previously redefined the Jews as "subjects" without the full rights of German citizens.[339] This subtle measure provided a very effective mechanism for persecuting "non-Aryans." Henceforth laws and decrees needed merely to assign certain rights exclusively to citizens in order to take them away from the Jewish subjects.

Finally, and as expected, the remaining "non-Aryan" full members had to leave the academy. Furthermore, the PAW was supposed to persuade these few Jewish members to resign quietly. The contrast between these final purges and the earlier Einstein affair is stark. Whereas in 1933 REM wanted to generate publicity for getting rid of Einstein, the ministry now did not want to draw attention to the fact that it had tolerated Jews in the PAW for so long. However, the National Socialist leadership did make a concession for the moment with regard to the foreign members: REM would not require that external and corresponding members satisfy the same requirements. Finally, REM gave PAW less than a month to report back to Rust.[340]

The academy membership and leadership capitulated immediately. The "non-Aryan" members were informed of Rust's decree by unofficial and confidential letters. The three scholars, Adolf Goldschmidt, Eduard Norden, and Issai Schnur, responded by resigning from the PAW. When the chairman reported this to the general meeting of the academy, he requested and received permission on behalf of the academy to express thanks to their former colleagues for their many years of valuable work. The PAW immediately began altering the statutes as ordered.[341] On 14 October 1938 the academy reported to REM that its Jewish members had left the PAW.[342] Thus the academy had purged itself of its last Jewish members before the infamous pogrom dubbed the "Night of Broken Glass" and the radical escalation of anti-Semitic terror and anti-Jewish legislation that followed.

For the Jews in Germany, 1936 and 1937 were relatively calm years in large part because the Third Reich wanted to present a good image for the 1936 Olympic games. But that changed dramatically in 1938. In the night of 9 November 1938 a murderous pogrom was unleashed by Minister of Propaganda Josef Goebbels, ostensibly in response to the assassination of a German diplomat in Paris by a Polish Jew. Throughout Germany, SS and SA (not in uniform) burned synagogues, destroyed seven thousand businesses, killed 100 Jews, and sadistically tortured thousands more. There were 20,000 Jewish men arrested and sent to concentration camps. Most Germans were shocked by the pogrom. Many people privately complained about the vandalism, lawlessness, and destruction of property. However, there was little or no opposition to the legal measures that followed. The National Socialists used the "Night of Broken Glass" as a cynical excuse for far-reaching decrees against the Jews, thereby excluding them from the economy and removing most, if not all, of their remaining freedom.[343]

The academy also felt the change in official policy towards Jews. Without warning in late November, REM specified additional changes in the new statutes. Members who were half-Jews, who had some Jewish ancestry, or who had Jewish wives had to leave the academy as well. Indeed, they were to be handled exactly as PAW had treated their full Jewish members. Rust considered exceptions inappropriate. Thus the National Socialists used an obvious yet effective tactic: no mention was made of the intention to get rid of the "half-Jews" until the full Jews were gone. Although the PAW was confronted with a series of escalating demands, each was presented as if it was the last and final concession and gave no hint of further measures to come.

Since only Reich citizens could become full members, in the future no Jews would be elected. Furthermore, the same standards would of course be used for the election of new corresponding or honorary members. In particular, REM would reject the election of a foreign member if he was a Jew in the sense of the Reich Citizenship Law. Existing corresponding and honorary Jewish members living in Germany would be asked to resign. If they refused, then

Rust would take advantage of the power given him by the new statutes and dismiss them. Finally, REM would postpone further action on Jewish foreign corresponding members until it had discussed the matter with the German Foreign Office.

Henceforth Rust would appoint the academy president, vice-president, and two secretaries, although the PAW was free to make suggestions. In order to rejuvenate the academy, full members over the age of seventy could be relieved of their duties, making possible the election of a younger full member. This apparent reform was a transparent method of silencing several recalcitrant older members and replacing them with younger scholars more congenial to National Socialism. Finally, REM asked the PAW to consider changing its name to "Berlin Academy of Sciences." As usual, the PAW had only a month to submit the new statutes to Rust.[344]

The external pressure on the PAW to transform itself was complemented by agitation by the National Socialist fifth column within the academy. On 1 December Vahlen, Bieberbach, and three other NSDAP members confronted the PAW leadership. These party comrades told their colleagues in the PAW that they had heaved a sigh of relief when REM demanded new statutes for the academy. Indeed, Vahlen, Bieberbach, and the others had felt ashamed that the academy had remained silent and not already voluntarily done what was needed. In other words, the academy should have voluntarily transformed itself into a completely "Aryan," National Socialist institution rather than waiting for the Ministry to force them to do so.

However, the five party comrades noted that a new epoch in the history of the PAW was beginning. They disagreed fundamentally with the argument made by many academy members that the PAW had to save what could be saved. Bieberbach and Vahlen reminded their colleagues that they all had been living since 1933 in a National Socialist state, where everything was to be arranged according to fascist principles, including science. It was not a matter of saving something, they argued, rather of building something new and National Socialist.

Since party comrades were best suited for such work, Vahlen, Bieberbach, and the others demanded that they be included in the committee charged with changing the statutes.[345] After a short discussion and one substitution, the academy agreed.[346] At the same meeting the PAW also capitulated to the demand that all members with some Jewish ancestry leave the academy. Acting chairman Planck read the REM decree requiring the removal of the members who were "part-Jewish or had part-Jewish wives" to the meeting. He then requested and received the permission to thank these members on behalf of the academy for their valuable contribution to the scientific work of the academy.

Implementation of the decree was entrusted to the statutes commission, now dominated by Bieberbach and Vahlen.[347] The academy did risk one pathetic request: that REM not apply this policy as strictly as had been done in the universities. Apparently some academy members still clung to the delusion that the National Socialist state would grant exceptions for Jewish members. Shortly before Christmas, the academy learned that the PAW members who were part Jewish or had part-Jewish wives, Max Sering, Otto Hintze, and corresponding members Felix Jacoby and Hans Horst Meyer, had resigned.[348]

On 22 December, PAW officially submitted its new statutes, which corresponded completely with the REM decree. However, the academy cautiously declined the suggestion of renaming the academy because the title "Prussian Academy of Sciences" was so well-known internationally.[349] The new statutes created the position of academy president, and the Ministry of Education immediately named Vahlen acting president.[350] A few weeks later Bieberbach was appointed acting secretary of the scientific class. When an academy member complained that the four academy secretaries had resigned their offices and cleared the way for Vahlen and Bieberbach without informing the academy and thereby forestalling any discussion, he was told that there had not been enough time.[351] This was either an excuse or the result of the tactics skillfully employed by REM to seize control of the PAW.

Thus the leadership principle was finally introduced to the academy in 1938 on the eve of World War II, a few months after the brutal pogrom of Germany's Jews and in the same year when Hitler purged the leadership of the armed forces. Conservative generals who had been critical of Hitler's foreign policy were forced to resign and replaced by more pliable men. The traditional German elites lost what little remaining autonomy they had within the National Socialist state. Now nothing stood in the way of Hitler's war.[352]

When the scientific class met on 19 January 1939, acting secretary Bieberbach announced that they had five free positions as replacements for older members. First, Bieberbach pointedly noted that he did not want to elect other relatively old scientists, rather the academy should bring in suitable younger colleagues. Here "suitable" had a specific meaning. Racial acceptability was now taken for granted. These new appointments had to meet an especially high standard with respect to political desirability, i.e., not merely being politically harmless, rather having special political qualifications or backing. However, Bieberbach artfully passed the buck. Neither he nor acting president Vahlen would make such a decision; that would be up to the responsible political offices.

The secretary went on in the January meeting to develop what must have been a deceptively seductive argument: of course, political qualifications would not replace scientific performance. Bieberbach assured his colleagues both personally and in the name of Vahlen that no one would be prepared to support the election of a member who did not completely and entirely fulfill the usual scientific requirements. In short, Bieberbach and Vahlen wanted only to require especially high political qualifications while maintaining the usual scientific standards.

In fact, there was no shortage of qualified scientists who also met these special political qualifications. Many of Germany's best scientists actively or passively supported National Socialist policies. Moreover, Bieberbach had been met with understanding from the political officials when he had argued to them that high scientific qualifications were an absolute prerequisite of any election. Bieberbach had taken the liberty of preparing a list of suitable names for new academy members, but assured his colleagues that he was prepared to discard any name for whom the representatives of the discipline had objections with regard to the scientific qualifications. In contrast, the mathematician did not offer to include any additional names in the list.

Next Bieberbach brought up the case of the physical chemist Max Volmer, yet he was not named specifically.[353] Although the academy had previously nominated him, representatives of the National Socialist state had found his political conduct unacceptable.[354] Thus Bieberbach drove home the point that he and Vahlen had not invented the high political standards for new academy members. That had been done by National Socialist officials in REM. What had happened with Volmer had been very unpleasant, and the academy had to avoid such situations in the future. Here Bieberbach and Vahlen also began another effective tactic: telling their colleagues—whether true or not—that the two of them had barely managed with great effort to keep the political authorities from punishing the academy for some matter or, even worse, from restricting the freedom of the PAW even further.

However, Bieberbach probably revealed his hypocrisy when he moved on to the next order of business: electing the future National Socialist Armaments Minister, Fritz Todt, as an honorary member of the academy. After arguing (rather implausibly) that Todt's scientific achievement matched that of the other honorary members and his political and economic significance for the German people far outweighed them, Bieberbach not only called upon his colleagues to elect him, he broadly hinted that any black spheres might cause problems for the academy. Todt was nominated with only a few votes against him.[355] A week later the full academy nominated Todt by a similar margin.[356]

However, the National Socialist leadership of the PAW had not yet won over their colleagues. Vahlen closed an academy meeting in late January with a personal and serious appeal to the members. They had to put aside their personal resentments, jealousies, friendships or antagonisms, he urged, in order to place the good of the whole above that of the individual. The time had come for camaraderie and support oí the acting leaders. Otherwise, Vahlen noted menacingly, the academy might suffer heavy damage.[357]

A few months later Vahlen turned his attention to the traditional secret ballot. The academy president noted that black spheres had repeatedly been deposited without any member having openly expressed his objections. This result is hardly surprising. It had always been common for a candidate to receive a few black spheres, and few members wanted to oppose openly a candidate backed by the PAW leadership. The implication of Vahlen's remarks was clear. The academy members could continue to enjoy their traditional secret ballots only if they always voted yes. The academy responded by electing twenty-four members *en masse*.[358]

Vahlen's increasingly dictatorial handling of the academy led to a modest revolt. Three senior academy members, Planck, Heinrich Lüders, and Hans Stille, criticized Vahlen's actions in writing and sent copies of their letter to all academy members. The acting president reacted by accusing his critics of unfairly mistrusting and trying to pressure him. Since Planck and his colleagues were hardly in a position to threaten Vahlen, the mathematician's response suggests that he was either concerned about his scientific reputation, or senility was causing him to lose his grip on reality.

Vahlen brought up the matter of confidence before the entire PAW and challenged anyone to discuss the supposed uneasiness among the members which had led to mistrust of the acting president. Planck now backed down and argued that the letter should not be seen as a statement of mistrust, rather they had merely expressed their concern for the future of the academy. The physicist went on to say that, in his opinion, the academy should have full confidence in Vahlen and be thankful for his efforts on its behalf. Vahlen was pleased to note that no one had expressed mistrust in him or the other academy officers, and thanked them for their support.[359]

Vahlen's acting presidency was due to run out on 15 June. When the academy met that day, the members were informed that Rust had accepted the new statutes with a few minor changes and that the PAW now had to nominate a new president, vice-president, and two class secretaries. Not surprisingly, Vahlen suggested that the academy vote on the four offices as a bloc, that is, they should vote to make the acting officials permanent. No doubt Vahlen hoped to avoid a referendum on his personal popularity.

But Planck stirred himself to raise a dissenting voice. In his opinion the academy president should be someone with very good connections to scientists in foreign countries and therefore could well represent the academy outside of Germany. Planck nominated Hans Stille as president. Another member supported Planck by noting that, even according to the new statutes, the PAW had to vote on its nominations for the four academy offices. Yet a third disagreed, and a long discussion with many participants followed.

Vahlen saw that an election was unavoidable, and called for a two-stage secret ballot for PAW president using slips of paper. The first round of voting determined the candidates and produced twenty-three votes for Vahlen, twenty-five for Stille, one each for Heymann and Planck, and five empty pieces of paper. The second round, now between Vahlen and Stille, ended in a tie.[360] The other three acting officers ran unopposed and were elected. Vahlen laconically noted that he would report these results to REM.[361] Two weeks later Minister Rust appointed the acting officers, including Vahlen, to their permanent positions.[362] The historian John L. Heilbron has characterized Planck's final challenge of Vahlen as a "moral victory" because Planck and the academy did not go down without a fight.[363] If so, it was one of the last such victories in the history of the PAW under Hitler.

✠   ✠   ✠

International Relations The PAW and other academies of science played an important role in the international commerce of science, often organizing or sponsoring conferences, corresponding with foreign institutions, and providing a forum where science policy on an international scale could be debated and created. Before World War I, German science dominated the international scientific community, German was the main language of science, and the PAW played a decisive role in the international politics of science.

When Germany lost World War I, the victorious allies imposed the Treaty of Versailles, a peace settlement which forced Germany to give up large amounts territory, to restrict its military, and to pay large reparations to some of the victors. Many Germans considered the treaty unfair and punitive, especially because of the war guilt clause which forced Germany to accept all blame for the war. Germany was now ostracized, and so was German science. In 1919 two new international scientific organizations, the International Research Council and the International Academic Union for the Humanities, were created in order to exclude Germany and Austria.

Many Allied scientists argued that time would have to pass and passions cool before they could reaccept their former enemies into the international community of science. The Germans simply considered it a boycott. This ostracism was fairly effective during the first postwar years. Congresses were not held in Germany, the German dominance of scientific journals was broken, and German was even replaced slowly by English. But the boycott had never been complete, and by 1925 it was beginning to crumble. By the late twenties many scientists in the United States and Europe wanted to reopen the channels of scientific cooperation. However, when the former allies became willing to accept the Germans, the latter began playing hard to get.

For the German scientists, the boycott was a moral issue. Their pride had been wounded. They tried to put up a united front and condemned the few deserters like Einstein, scientists who accepted personal invitations to attend conferences when Germans officially were banned or at least unwelcome. When German foreign policy changed in the course of the Weimar Republic from confrontation to cooperation with the League of Nations and the German Foreign Office turned to German scientists for assistance in reestablishing international ties, the German scientific community refused to cooperate. When Germany was invited to join the International Research Council in 1926, the cartel of German academies and Union of German Universities refused. It quickly became obvious that they simply did not want to join this organization, in large part because of the bitterness caused by the boycott.[364]

Although the PAW refused to participate in international scientific activities coordinated by the Council, it did take an active part in National Socialist cultural policy. In late June 1937 REM asked the academy if it was able and willing to name foreigners or Germans living outside of Germany who were actively working for German interests as honorary or corresponding members for the sake of cultural and political considerations. The academy was willing, with two conditions: the individual must fulfill the academy's usual scientific requirements and the relevant experts must be willing to propose him.[365] This was the same bargain that Bieberbach had offered with respect to full membership. The PAW was willing to bestow scientific honors for political reasons, so long as they went to good scientists.

The National Socialist government closely monitored and controlled the international activities of academy scientists. For example, in the summer of 1938 REM informed PAW that all invitations to an international medical congress in Strasbourg were to be turned down, perhaps because Germany had been forced to return Alsace to France as part of the peace settlement.[366] In late October PAW received an invitation to attend a congress on cancer research in Paris. Since REM considered German participation undesirable, PAW turned down the invitation with thanks.[367] However, scientists were welcome to get involved in politics, so long as it suited National Socialist interests. Shortly after Germany had absorbed Austria and took one of the first major steps towards World War II, Walther Nernst suggested that the Berlin academy send a telegram of greetings to the Vienna academy and welcome them home to the Reich. His colleagues agreed.[368]

The successful German Lightning War (*Blitzkrieg*) radically changed the quality of the PAW's international relations. It was no longer a matter of whether German academies would cooperate with international organizations in Belgium and France, rather what the conquering Germans would do with them and the rest of occupied Europe. War also brought with it additional financial restrictions. Vahlen announced that the academy finances were being reevaluated and that until further notice the academy would not publish the work of non-Germans.[369] But exceptions were made. In April Bieberbach successfully argued that the work of a Bulgarian mathematician should be published because the work was of high quality and it would be good for Germany's cultural relations with Bulgaria.[370]

The 1939 German–Soviet Non-Aggression Pact on the eve of World War II surprised and dismayed Germany's neighbors. The two totalitarian states set aside their deep ideological differences, agreed not to attack each other, and in a secret clause of the treaty divided Poland and the Baltic states between them. Hitler wanted the treaty so that his back would be free when he attacked western Europe, even though he intended to attack the Soviet Union eventually. Stalin wanted more time to prepare for the confrontation with Germany, because he in turn considered German aggression inevitable.

This pact also caused a radical about-face in official cultural policy. Cooperation between the PAW and Soviet institutions had previously been tightly controlled. In December 1936 the Reich Exchange Office in the Prussian State Library, which controlled and coordinated all exchanges of publications with foreign institutions, had ordered the PAW to provide them with a detailed list of every exchange with the Soviet Union, and informed the academy that any new exchanges would have to be approved in advance.[371]

Three years later, cooperation with Soviet institutions was positively encouraged. REM decreed on 30 November that scientific relations with the Soviet Union would be renewed.[372] By the new year the PAW was able to report that the previous exchanges of publications between the PAW and the Soviet scientific institutes had been reinstated, along with many new requests for German publications. The academy tried as best it could to fulfill the many requests.[373] The situation changed abruptly once again in the summer of 1941, when Germany tore up its pact and invaded the Soviet Union.

War had an immediate effect on the academy's international communication. Almost no exchanges remained with hostile countries. Allies were another matter. In November 1940 two more requests for publication exchanges from friendly countries were approved: a geophysical institute in Italy and a mathematical institute in Japan.[374] Countries that had been conquered by Germany offered special opportunities for international scientific cooperation. REM instructed Bieberbach to support an "Analytical Bulletin" being published by the "National Center for Scientific Research and Documentation" in Paris. This bulletin provided brief summaries of the contents of articles from scientific and technical journals from around the world, and was designed to facilitate the absorption of French industry by its German counterpart by encouraging the French to collaborate with the Germans.[375]

The academy also took part directly in the plunder of European science. In the summer of 1940 the Prussian State Library informed the PAW that manuscripts and library material of German origin were being returned, i.e., removed, from French and Belgian libraries. Furthermore, the academy was encouraged to place orders for such material.[376] In fact the PAW did order materials from libraries in occupied countries and thereby participated directly in the German rape of Europe and fundamentally perverted the purpose of an academy of sciences. This ruthless collaboration with National Socialism was an ironic twist on the academy's traditional fostering of international cooperation in science through an exchange of publications.

Perhaps the most consequential role played by the PAW in the cultural exploitation of countries under German occupation came in occupied Poland.[377] In late August 1940 the PAW informed the director of the university library in Berlin that there were nineteen publications of the Krakow academy which the PAW did not have. The PAW asked this official to arrange that these publications be sent to the Berlin academy from the former Krakow academy, which had been closed by German officials.[378]

In November the PAW was contacted by the newly established State and University Library in Posen, also in occupied Polish territory. The librarian was trying to build up a German-language library, and hoped to receive PAW publications. Although the new Reich University of Posen had only just been established and the librarian did not have much German literature to offer in exchange, he did have large collections of Polish literature which he would be willing to send to Berlin. The PAW responded immediately that it would be pleased to begin a publication exchange. It would send its usual publications to Posen together with a list of one German and eleven Polish publications which it would like in return.[379]

Sometimes this process was pushed by higher officials as part of the German policy of assimilation. When Education Minister Rust visited Posen, he pointedly noted that its library only had PAW publications through World War I. REM instructed the academy to send the missing publications to Posen.[380] Shortly after the new year the Posen library sent eighteen volumes to the PAW.[381] A few weeks later, the German occupation government in Poland sent PAW volumes from the archives of the former Polish Academy of Sciences.[382] The PAW also elected the rector of the University in Posen a corresponding member of the academy in 1941.[383]

The Polish scientists and scholars had little opportunity to protest the plunder of their country, but the special National Socialist brand of international scientific cooperation was not always passed over in silence. Early in December 1943 the PAW and the other German academies received a polite yet accusatory letter from the Swedish Academy of Sciences. Sweden was one of the few neutral countries during World War II. The German authorities in Norway had responded to student protests by closing the University of Oslo, arresting the male Norwegian students along with many teachers, and announcing that they would be deported to Germany for forced labor. How, the Swedes asked, did the PAW justify this?[384] The PAW's first reaction was to do nothing before first checking with the Foreign Office.[385] REM forbade both official and personal responses by any academy member.[386] One scholar nevertheless disobeyed the ministry and answered his Swedish colleagues in the spring of 1944 by reciting a list of destruction done to German culture by Allied bombs.[387]

✠   ✠   ✠

Vahlen's Presidency Vahlen and Bieberbach stuck to their promise and only elected competent scientists as full members, some of whom were National Socialists, some who were politically useful, and others who were politically harmless. In June 1939 Otto Hahn and two colleagues proposed Adolf Thiessen for full membership in the academy.[388] Thiessen was a capable scientist and long-standing National Socialist who had taken over the old institute of the Jewish physical-chemist Fritz Haber, after the institute had been purged of its "non-Aryan" scientists and Haber had been driven into exile.[389]

In 1943, Werner Heisenberg and Otmar Freiherr von Verschuer were elected unanimously to the academy.[390] Verschuer certainly fit the image of a "Nazi" scientist. He was the mentor of Josef Mengele and carried out research with the remains of concentration camp victims which his former student sent him from Auschwitz. However, Heisenberg's election demonstrates that another type of scientist was also acceptable to the academy: an apolitical scientist who was nevertheless considered valuable by the National Socialist state. In a modest act of defiance the scientific class voted in early March 1941 to nominate Volmer once again as a member.[391] However, the academy leadership simply ignored them.

By 1939 the PAW was completely integrated into the National Socialist state. In late February the PAW finally eliminated voting by spheres in favor of what was cynically described as free and open voting.[392] REM and the PAW also continued their relentless expulsion of "non-Aryan" members. In the summer of 1939 Ernst Heymann informed the academy that the Jewish scientist Richard Willstätter had been expelled as a corresponding member of the PAW. Willstätter had merely been informed that, according to the new statutes he was no longer a corresponding member because he did not fulfill the requirements for the Reich citizenship. No member of the academy raised any objections.[393]

In November 1941 REM informed the academy that the Italian corresponding member Tullio Levi-Civita was a full Jew. Vahlen noted that he must now be removed from the list of corresponding members, and the academy moved to do so. The physicist and corresponding member James Franck was supposed to be a full Jew, but since he was in the United States, REM decreed that a decision in his case would have to wait until after the war.[394] However, once Germany was at war with the United States the situation changed. In November 1942 both Franck and Max Born as "non-Aryans" were removed from the list of corresponding members.[395]

Although in a sense Vahlen had now reached the zenith of his power within the PAW, the start of World War II revealed that his mental facilities were deteriorating rapidly. In October he requested that REM transfer him to a position where he could actively contribute to the war effort. In November 1939, the septuagenarian mathematician informed the personnel office of the SS that he was available if the personal protection of the Führer needed strengthening. A few months later he asked the same office for permission to wear a field gray uniform and for assignment to the front. But Vahlen's superiors kindly turned down his offers.[396]

In early 1943 Vahlen, who was clearly steadily losing his grip on reality, submitted his resignation to Rust in order to go to war. The matter was passed onto the SS, where Vahlen's colleagues tried to say no gently.[397] The Ahnenerbe, the SS scientific research branch, thought that Vahlen's offer was a nice gesture, knowing full well that Vahlen's faculties were not what they used to be. Of course, no one wanted to hurt Vahlen's feelings. Perhaps SS leader Heinrich Himmler could himself tactfully decline the offer.[398] In fact, on 25 March, Himmler told Vahlen that an old fighter like himself had nothing to prove. Instead, he should devote himself to his scientific research.[399]

Vahlen's memory began failing him so often that the academy business suffered, creating difficulties and embarrassing situations. The mathematician was finally relieved of his duties as academy president in the summer of 1943. But Himmler did promote Vahlen within the SS[400] and early in 1944 the SS finally gave Vahlen permission to wear a field gray uniform.[401] Vahlen tried one last time in February 1944 to join the Waffen-SS, the military arm of the SS. Once again, Vahlen was gently advised to devote himself to science.[402]

The war finally came home to the academy in the summer of 1941. The president began one meeting by honoring two former scientific employees of the PAW who had fallen on the eastern front.[403] In 1943 Allied bombing raids became common over Germany, causing death and destruction and revealing the impotence of Göring's air force. However, the raids did not have the hoped-for effect on morale. The more the Germans suffered, the more they stuck together and fought their enemies. Goebbels' propaganda now emphasized the "total war" and the atrocities which the Soviets would commit if they made it to Germany.

In late November several academy members lost everything they owned to Allied bombs.[404] By mid-December, the bombing had made further printing of the academy publications impossible for the duration of the war.[405] The first meeting of the academy in 1944 was held in the air raid shelter because the usual meeting room had been damaged.[406] The air raids and small number of members present ended the meeting on 9 March after just ten minutes.[407] Vahlen lost his apartment to an Allied air raid and moved to Vienna, where he was immediately given a honorary professorship by the Vienna Technical University.[408] In July 1944 Rust told the PAW that no new president would be named until after the war.[409] The last minutes of an academy meeting in the Third Reich noted merely that the meeting had to be postponed.[410]

✠   ✠   ✠

Postwar After the fall of the Third Reich, what was left of the PAW scrambled to accommodate itself to the new political realities. The academy began meeting again in June 1945, even though many members were no longer in Berlin and the city was occupied by Allied and in particular Soviet forces. The acting secretary, Hermann Grapow, told his colleagues that the local mayor was very interested in cultural matters and had offered to help find a permanent meeting place for them—they were now meeting in the local city hall. However, a different academy member, Eduard Spranger, objected to the apologetic tone of the draft report which was to be submitted to the authorities. In his opinion, there was no reason for the academy to begin apologizing for its former conduct.[411] He was soon proved wrong.

In mid-June PAW member Johannes Stroux met with the local magistrate about the financing of the academy, new statues, and office space. The magistrate asserted its veto power over the election of full members. When the academy subsequently discussed how the elections should take place, one member proposed voting by acclamation, a suggestion perhaps unconsciously reminiscent of the Third Reich. The rest of the members agreed that a secret vote using slips of paper was preferable.[412]

The PAW now took great care to ingratiate itself with the new rulers. When a local politician suggested that the academy start public lectures as a way to attract attention, the PAW responded by proposing a lecture on the connections between the writer Jacob Grimm and Russian scholars.[413] The academy also sent a congratulatory letter care of the Soviet occupation government to the Leningrad Academy on their 220th anniversary.[414] The Germans' concern about the future of their academy was justified. In a subsequent meeting with German officials employed by the Soviet occupation government, the academy representative was told that the government was not certain that the PAW still existed, rather it might have to be refounded. In other words, all of the existing members would in effect be dismissed and the academy rebuilt from scratch. This barely veiled threat was followed by the pointed remark that the PAW still employed former members of both the NSDAP and SA.[415]

At the next meeting, in July 1945, the academy discussed what to do about their colleagues who had been members of the NSDAP, forcing five former party members to leave the room. Many more employees of the academy had been in the party, and the remaining PAW members decided to dismiss four employees and try to keep two others. Since the city government refused to allow the academy to impose one policy on its employees but another on its members, the PAW could not delay dealing with its politically tainted members. The officials responsible for the PAW gave its acting president two lists, one of eight members who had to go, and another with the names of eighteen individuals who had to be examined more closely. The members present accepted the two lists and the proposed measures unanimously.[416] The list of eight included Vahlen, Bieberbach, Konrad Meyer, Peter Adolf Thiessen, Franz Koch, Carl August Emge, Friedrich Stieve, and Theodor Mayer.[417]

The PAW had to face criticism for its past under National Socialism and pressure to conform to the wishes of the Soviet Occupation Government. In August the shadow of the Einstein affair reemerged. The magistrate ordered the PAW to use its records to prepare a report on the entire matter.[418] The PAW was also informed that its library, like all libraries, would be purged of undesirable political writings.[419] In December the PAW began to transform itself once again, this time according to the model of the Soviet Academy of Sciences by incorporating scientific institutes. One of the first to be considered was a new institute for Slavic languages, but the PAW also began to swallow up various scientific state and KWG institutes in and around Berlin that had been orphaned by the division of Germany.[420] Thus the introduction of the Soviet model had both ideological and pragmatic justification.

Members also protested in vain against the dismissal of former NSDAP members and the grave dangers this policy had for the academy, the university, and indeed for the cultural life of Berlin in general. The result would be the emigration and persecution of respected scholars on one hand, and the obstruction and alienation of new forces on the other.[421] It is worth noting that even though this protest was futile, it went far beyond any criticism made by the academy of National Socialist policy in the early thirties.

Shortly after Christmas 1945 the PAW as such ceased to exist. After a sometimes bitter discussion the academy bowed to pressure to change its name—something not even the National Socialists had insisted upon—and remove the name Prussian. It was now the "Berlin Academy of Sciences."[422] By 1947 it had been renamed the "German Academy of Sciences" and included over nineteen institutes and other scientific institutions.[423] This academy in turn became the "Academy of Sciences of the German Democratic Republic" in 1972 and lasted in this form until German reunification in 1990, when the institutes were either disbanded or reconnected to some other scientific institution. What remained has returned to the original model for an academy of sciences, now renamed the "Berlin-Brandenburg Academy of Sciences."

On 5 October 1946, something else happened that was ironically reminiscent of the Third Reich. By order of the Soviet occupation government, all institutions under its control, including the academy, gathered together at a rally celebrating the judgment reached at the Nuremberg Trials. With few exceptions, the surviving National Socialist leadership was sentenced to death for crimes against humanity. The academy was ordered to ensure that all members and employees attended, and that they arrived in a group.[424] There were certainly grave differences between the Third Reich, now condemned for the crime of genocide, and the Stalinist society the Soviets imposed on eastern Germany. But the coerced public ceremony recognizing the judgment from Nuremberg is nevertheless reminiscent of the mandatory collective listening to Hitler's speeches during the Third Reich and illustrates the special tragedy of scientists and other Germans in the Soviet occupation zone and the subsequent German Democratic Republic. They traded a murderous racist dictatorship for a milder, socialist one.

The National Socialist scientist and PAW dictator Theodor Vahlen barely outlived the Third Reich. According to his widow, the seventy-six-year-old Vahlen died on 16 November 1945 in a Prague prison.[425] Ludwig Bieberbach, perhaps the other most prominent National Socialist in the academy, was one of the very few professors who never regained a teaching position in Germany. However, the postwar stigma attached to Bieberbach was mainly caused by his infamous *Deutsche Mathematik*, not because of his role in the subversion of the PAW.

The concessions Max Planck made during the Einstein affair were not forgotten, but have usually been softened by the emphasis placed both on the statement Planck made before the academy honoring Einstein's scientific achievement, and on the great personal suffering Planck had to endure under Hitler. The final and ultimately futile fight Planck put up for the independence of the academy demonstrates that he saw clearly what National Socialism was doing to the academy, to science, and to Germany.

Perhaps the strongest image associated with the PAW under Hitler is Max von Laue's barring the door to the "Nazi" physicist Johannes Stark. Von Laue himself published an account of the affair in 1947 as a response to a self-serving article from Stark. Indeed von Laue is regularly portrayed as one of the few German scientists who refused any and all compromise with the National Socialists, and the Stark affair is presented as proof. The history of the academy under Hitler demonstrates that his conduct in fact was more ambiguous and ambivalent, but nevertheless still laudable.

The PAW was important enough to be brought into line with the rest of German society during the Third Reich, but the slow pace of the transformation of the academy and the subsequent imposition of Vahlen as PAW president also reveal that the academy was really not that important to the National Socialist state. Otherwise its Jewish members would have been thrown out immediately, and it would hardly have been used as a rest home for senile party comrades. The PAW could delay its purge of "non-Aryan" members because it was relatively unimportant for National Socialist science policy, not because of the personal or professional courage of its members.

The academy certainly did not actively oppose or resist the new regime, but that does not necessarily earn it the "Nazi" label. On one hand, the academy began immediately to make concessions to Germany's new National Socialist rulers. On the other hand, with few exceptions the PAW continued throughout the Third Reich to have outstanding scientists and scholars as its members who produced high-quality science and scholarship. The members of the PAW willingly and knowingly cooperated with National Socialist policies while simultaneously trying to maximize their shrinking professional and personal independence.

Bieberbach's and Vahlen's argument, that only good scientists would be chosen for membership, even if they also had to fulfill political criteria as well, was no doubt both seductive and effective. Any member who wished to believe that the academy was apolitical and that scientific qualifications were all that mattered could accept this perverse type of affirmative action for professionally competent National Socialist scientists and their fellow travelers.

What such a scientist could not do, however, was to dwell for too long on those scientists who had been driven out of the academy or who were denied admission. As the political scientist Joseph Haberer recognized, compliance and cooperation did not protect the Academy, rather helped transform it into a willing tool of National Socialism. Furthermore, in the long run, the unwillingness to protect colleagues and the concessions made to the regime were the most grave legacy of this period.[426]

The history of the academy shows that its members were cajoled, coerced, threatened, and seduced step by step into transforming themselves into a willing tool of the National Socialist state. This transformation culminated in the ruthless purge of "non-Aryan" members and participation in the scientific rape of occupied Europe. But Vahlen did not conquer and subsequently pervert the academy, rather he took over after its members had already collectively sealed a Faustian pact with the Third Reich. Bieberbach did not undermine the academy by himself, rather he was able to persuade the majority of his colleagues to either help or at least not oppose him. Planck not only regretted the shameful handling of Einstein, he also was forced to preside over the forced resignation of the rest of his Jewish colleagues. Finally, von Laue did bar the door to Johannes Stark, but he and Planck also had little choice but to step aside when others held the door open to scientists like Fischer, Thiessen, von Verschuer, and Vahlen.

6

Physics and Propaganda

The majority of German scientists neither embraced National Socialism nor emigrated from it. They stayed and worked, either withdrawing as much as possible from the disturbing reality of the Third Reich—often called "inner emigration"—or actively participating in the National Socialist system. The latter individuals inevitably acted in an ambiguous and ambivalent manner. Enthusiastic National Socialists, opponents, opportunists, and the vast silent majority all worked within the system despite having very different motives. Thus different observers have often described the same activity by the same scientist either as collaboration or resistance. Both labels are problematic because they mirror the black-and-white juxtaposition of "Nazi" and "anti-Nazi." For most scientists, the day-to-day reality lay in between.

Werner Heisenberg's guest lectures in foreign countries and resulting participation in cultural propaganda during the Third Reich provide an excellent example of how ambiguous and ambivalent cooperation with the National Socialist state could be. One lecture in particular deserves close inspection. In September 1941, when German armies were pushing deep into Soviet territory, Heisenberg and Carl Friedrich von Weizsäcker traveled from Germany to Copenhagen, where they gave talks and visited their Danish physicist colleagues. This visit remains one of the most controversial events in the recent history of German science precisely because it has been used as evidence for diametrically opposed interpretations of Heisenberg's and von Weizsäcker's conduct under Hitler: (1) they went to Copenhagen in order to help their colleague Niels Bohr and to save the world from nuclear weapons; (2) they went in order to help the National Socialists exploit Bohr (who had a Jewish mother) and win the race to the atom bomb.[427]

Heisenberg's and von Weizsäcker's 1941 visit to Denmark belongs in the context of National Socialist cultural propaganda in countries occupied by or obedient to Germany during the war. The physicists did not simply go to Copenhagen to help Bohr. They traveled to a Denmark occupied by German troops. While in Copenhagen, they participated in official propaganda by lecturing at a German cultural institute.

Heisenberg's many guest lectures also facilitate an analysis of two important aspects of science during the Third Reich. First, the National Socialist regime transformed foreign lectures and international conferences into effective tools for cultural propaganda. Second, there was a functional relationship between the changing official attitude towards Heisenberg, the rehabilitation of modern physics under Hitler, and the usefulness to the National Socialists of Heisenberg as a goodwill ambassador.

Perhaps most important, Heisenberg's and von Weizsäcker's September 1941 trip to Copenhagen must be placed in the context of World War II. For this reason the story of their foreign lectures will be divided up into two chapters. Chapter 6, "Physics and Propaganda" covers the prewar period and the Lightning War, when it appeared that the war would soon end with a National Socialist victory. In contrast, Chapter 7, "Goodwill Ambassadors" covers the period when the war turned sour for Germany and the persecution of the Jews was transformed into the Holocaust.

✠   ✠   ✠

The "Coordination" of Foreign Lectures The National Socialists took care to regulate quickly and strictly any cultural exchange with other countries as part of a thorough "coordination" of the civil service. Officials at REM informed the rectors of the German universities that they welcomed foreign lectures by German scientists, so long as the scholar was worthy of representing Germany in the National Socialist sense. Only REM could approve a foreign trip by a civil servant or employee under its jurisdiction, which included all university instructors.[428] By early 1934 the ministry noted in a threatening tone that individuals with unsuitable personalities and ideologies were being proposed as representatives of the new Germany.

All foreign travel requests for speaking engagements were to be submitted through official channels and had to include and quote verbatim the opinion of the regional leader of the NSDAP.[429] The Foreign Office of the new Germany now demanded that it be informed ahead of time of any foreign lectures, and that the speaker contact and work closely with the German Embassy in the country to be visited.[430] Moreover, this strict policy was introduced at a time when the Foreign Office was still relatively independent of National Socialist influence.

By early 1935, REM had extended its control to lectures by foreign scholars inside of Germany. Any invitation had to be approved by the ministry in advance, and any such request had to be submitted early enough so that the ministry could check with both the Foreign Office and the German Embassy in the country concerned. The Education Ministry also extended its right of refusal. As of June 1935 no invitation either for a lecture abroad or for participation in an international congress could be accepted without its permission.[431] By 1937 the Ministry of Education required that universities and scholars provide complete information on all conferences being planned, both inside and outside of the Reich.[432] The Ministry of Propaganda also gained some control over international cultural commerce. Its German Congress Center controlled the technical aspects of such trips by providing the scholars with foreign currency and through the organization of congresses held inside Germany.[433]

Differences in the treatment of nationalities under these guidelines illustrate how sensitive cultural policy was to political events. In 1935 Germans living abroad could be invited to purely scientific conferences in Germany without consulting REM, but any visits to or from Poland or Alsace-Lorraine had to be approved well ahead of time. In the spring of 1936, the Education Ministry forbade German scholars to have anything to do with any organization or event connected with the League of Nations. By that October, all official visits to Spain by civil servants were to be cleared beforehand. A month later this decree was extended to cover all employees.[434]

In 1927 the twenty-six-year-old Heisenberg was called to a full professorship in theoretical physics at the University of Leipzig. Heisenberg, one of the creators of quantum mechanics, quickly received honors, recognition, and invitations from abroad. In 1929, Heisenberg was invited to hold a series of guest lectures at the University of Chicago during the summer semester.[435] Three years later Heisenberg was granted leave again to lecture at a summer school for physics at the University of Michigan. Heisenberg's guest lectures continued after the National Socialists took power, but the context in which these goodwill trips took place became very different.[436]

The year 1933, which included such radical change in Germany, also brought good news to Heisenberg in the form of the 1932 Nobel Prize for physics. The University of Leipzig was very proud of Heisenberg, but concerned that he might now be tempted to go elsewhere. Heisenberg responded with thanks for the appreciation, noted that the philosophical faculty had made his stay in Leipzig as pleasant as possible, and that he hoped to be able to stay at the university for a long time to come.[437] In the spring of 1934 Heisenberg received a call to a position at Harvard University with very generous fringe benefits. When Heisenberg informed the dean of this American offer, the administrator in turn assured Heisenberg that he would spare no effort to try and retain the physicist for the University of Leipzig and Germany. The Nobel laureate decided to stay at Leipzig, at least for the time being.[438]

In February 1936 Heisenberg requested another leave of absence to lecture at the University of Michigan in July and August, and to attend the tercentennial anniversary celebrations of Harvard University. The ministry approved Heisenberg's trips, granted him leave from July through September, and informed the Foreign Office and the Congress Center of his plans.[439] In May he submitted an application to attend a physics conference at Niels Bohr's Institute for Theoretical Physics in Copenhagen, which was approved as well.[440] In his subsequent report on the conference for the ministry, Heisenberg restricted his comments to scientific matters and avoided politics. In contrast, Pascal Jordan, another of the creators of quantum mechanics but an enthusiastic follower of Hitler,[441] submitted a report couched in National Socialist rhetoric.

In the spring of 1937, Heisenberg requested permission to attend a congress on statistics to be held that October in Geneva. He had been invited to deliver one of the featured papers, a lecture on "Statements of probability in the quantum theory of wave fields."[442] The rector approved the trip, but the local head of the University Teachers League was ambivalent.[443] Although Heisenberg had never been a radical leftist, had always been nationalistic, and had volunteered for military training the previous autumn, the Party official had some misgivings about approving the trip to Switzerland. Heisenberg had close connections with Jewish physicists in foreign countries and, apparently worst of all, rejected anti-Semitism. One could not expect that Heisenberg would represent National Socialist doctrine while outside of Germany.

But despite these misgivings, the University Teachers League approved the trip because of Heisenberg's international reputation. He was so well known, inside and outside of Germany, that the prestige of the National Socialist government would be hurt more by denying him the chance to travel to Switzerland than by

Niels Bohr (right) and Werner Heisenberg in the Niels Bohr Institute, Copenhagen, 1936. (Photo by P. Ehrenfest, Jr., Courtesy of the AIP Emilio Segrè Visual Archives.)

giving him permission for the trip.[444] One probable reason for this ambivalence was the fact that public political attacks on Heisenberg had begun, for example in the main newspaper of the NSDAP, the *Völkischer Beobachter*.[445] It is not clear whether Heisenberg went to Geneva or not. When Heisenberg requested permission in the summer of 1937 to go to the annual small conference at Niels Bohr's institute in Copenhagen, no objections were raised.[446] Perhaps Switzerland was considered politically more sensitive than Denmark, or the fact that Heisenberg went to Copenhagen so often made the trip seem less dangerous.

Events surrounding a nuclear physics conference held in Zurich in the summer of 1936, which Heisenberg could not attend since he was in the United States, are instructive of the development of National Socialist cultural policy. Eight physicists asked for permission to attend the meeting, and six applications were approved. For Ludwig Bewilogua, Robert Döpel, Hans Geiger, Gerhard Hoffmann, and Fritz Kirchner, the ministry approved easily, if not swiftly.[447] Hans Geiger submitted his request on 23 May, and on 17 June had to write his rector again to accelerate the process. Geiger was scheduled to give the featured lecture in his own special field of research. It was in the interest of German science, Geiger argued, that he be allowed to attend, otherwise a Dutchman or a Frenchman would take his place.[448]

Rausch von Traubenberg, a professor at the University of Kiel with a Jewish spouse, ran into political trouble. The rector, the dean, and the representative of the University Teachers League, the Party organization in charge of university instructors, all approved the trip. The rector said that he could not imagine any serious danger in sending Traubenberg to the conference, which was to be limited to scientific matters. But Traubenberg had failed in the past to get permission to travel. The regional Party leadership of the state Schleswig-Holstein had killed all previous applications, and refused yet again.[449] The Reich Ministry of Education told the rector at Kiel to inform Traubenberg that he could not go to Zurich because of the shortage of foreign currency.

Fritz Sauter, who taught physics at the University of Göttingen, and in 1939 joined the NSDAP, submitted his request to attend the Zurich meeting, and as far as he knew, it went through without any problem.[450] In fact, REM approved the trip, only to learn that Sauter was being watched by the Gestapo, the domestic secret police branch of the SS. The ministry did not want to take responsibility for sending Sauter under these circumstances to Switzerland. Officials from the ministry then reached a compromise with the secret police.[451] Sauter could go to Zurich, but he would have to submit a report to REM on the attitude of Swiss physicists toward the new Germany.[452]

The request of Erich Regener, a physicist at the Technical University in Stuttgart, was forwarded on to the ministry with an unofficial letter that implied that Regener and his wife were not "Aryan."[453] REM responded by asking the Württemberg Ministry of Culture whether Regener had submitted the questionnaire required of all civil servants, and in particular, whether Regener had ever belonged to a Freemason Lodge and whether evidence had been presented that Regener and his wife were "Aryan." The Reich official made clear that, if at all possible, this information should be gathered without Regener's knowledge.[454] The Württemberg Ministry responded that Regener had never belonged to a Lodge and was of "Aryan" blood. His wife was Jewish.[455] A few weeks later, REM directed the Minister of Culture in Stuttgart to inform Regener that his trip could not be approved because of the shortage of foreign currency.[456]

✠ ✠ ✠

The "White Jew" and "Ossietzky of Physics" The National Socialist regime went to considerable lengths in 1935 and 1936 to present its best face, for example during the 1936 Olympic games. But during the last few years before the war, the more radical and disturbing aspects of the new Germany emerged, including the pogrom known as the "Night of Broken Glass" and the aggressive German military expansion.[457] These years were also very hard on Heisenberg. He suffered political attacks that were not only dangerous in themselves, but injurious to his personal and professional pride.

On 15 July 1937 he was attacked as a "white Jew" and "Jewish in spirit" by his colleague, fellow Nobel laureate, and president of the Imperial Physical-Technical Institute, Johannes Stark, in an article published in the SS weekly *Das Schwarze Korps*.[458] Heisenberg called upon his superiors to protect him against Stark's attacks. A fundamental decision was necessary. If the ministry considered Stark's viewpoint in *Das Schwarze Korps* correct, then Heisenberg would resign; if the ministry did not support such attacks, then Heisenberg demanded the sort of protection which the armed forces would grant to its youngest lieutenant. Heisenberg suggested that perhaps the Leipzig University Student Organization could do something, since it was affiliated with the NSDAP. He apparently thought that he had National Socialist allies in Leipzig.[459]

The bureaucracy did not welcome Stark's attack. Very many individuals lost their positions or were denied promotions on political grounds during the Third Reich. But the Ministry of Interior insisted that such decisions as well as any complaints about the political reliability of civil servants go through official channels.[460] Both the Ministry of Propaganda and the Party Chancellery had decreed in 1936 that attacks on civil servants in the press should be avoided.[461] The rector of the University of Leipzig—who brought the matter to the attention of the Reich regional representative in Saxony—observed that Stark had implicitly criticized those parts of the National Socialist government responsible for personnel policy and requested that the government enforce its policy towards such attacks in the press.[462]

Heisenberg continued his aggressive tone with his superiors. Almost seven months after he had insisted on either resignation or protection, he demanded to know whether the ministry believed that his performance deserved insults like "white Jew" and the "Ossietzky of physics"? Stark's attack and the inaction of his superiors had crippled Heisenberg's work. A student had turned down both a place and a stipend at Heisenberg's institute after Stark's attack out of fear that association with Heisenberg could harm him politically. This case showed that unless a clear decision was made concerning the attack in *Das Schwarze Korps*, work in Heisenberg's institute would be made almost impossible.[463] (The socialist and pacifist Carl von Ossietzky was awarded the Nobel Peace Prize while imprisoned in a German concentration camp, thereby embarrassing the National Socialist government and prompting Adolf Hitler to forbid German citizens thereafter to accept the Nobel Prize. Ossietzky died in the camp.)

Heisenberg also contacted the SS directly, but a low-ranking official informed him that they could do nothing for him. It appeared that SS Leader Heinrich Himmler and Minister of Education Bernhard Rust had decided not to answer Heisenberg's requests for the Munich professorship and for public recognition of his service and loyalty to the fatherland. Heisenberg saw no alternative but to submit his resignation at Leipzig and to leave Germany. He did not want to emigrate, he told his mentor Arnold Sommerfeld, but he also had no desire to live in Germany as a second-class citizen.[464]

Meanwhile, Johannes Stark had not prospered. He had refused as president of the German Research Foundation to fund some scientific research desired by the SS, and was subsequently sacked by the REM and replaced by SS man Rudolf Mentzel. In the spring of 1936 Adolf Wagner, one of the most powerful and ruthless regional party leaders in Germany, instituted legal proceedings to throw Stark out of the Party for having meddled in the politics of Wagner's region in southern Bavaria. Stark fought back and remained in the NSDAP, but his trial dragged on until 1938. After 1936, he was viewed with increasing disapproval within the SS and influential Party circles.[465]

Influential colleagues also intervened on Heisenberg's behalf. During the summer of 1938 the aeronautical engineer Ludwig Prandtl convinced Himmler that Germany could not afford to lose Heisenberg, who was still relatively young and could train a generation of scientists.[466] Prandtl was in a position to influence the SS. In 1937 the Party official in charge of Göttingen described him as a typical scientist in an ivory tower. Prandtl was an honorable, conscientious scholar from an older generation concerned with his integrity and respectability. However, given Prandtl's exceptionally valuable scientific contributions toward the expansion of the Air Force, he was also someone the National Socialists neither could do without, nor wanted to alienate.[467]

The leader of the SS forbade further political attacks on Heisenberg, invited the physicist to meet with him, and made it clear that he expected Heisenberg to stick to physics, not politics.[468] Heisenberg responded immediately, agreed to avoid politics, but insisted on a public rehabilitation.[469] In November a messenger from Himmler arrived and asked Heisenberg for more detailed information on the "physics war" between *Deutsche Physik* and the established physics community, which Heisenberg considered to be a good sign.[470] At the same time a Party official told Prandtl that the struggle against the theory of relativity had been stopped by someone in a high position.[471]

Despite its power, the SS could not end Heisenberg's troubles. In December 1938 an official from the Saxon Ministry of Culture paid an unofficial visit to his Berlin colleague in the Education Ministry and asked about the Heisenberg case. Minister Rust had not made up his mind, in part because the Heisenberg affair was only one part of the controversy between theoretical and experimental physics. The two bureaucrats agreed that Stark had gone too far. But they also agreed that Heisenberg had brought much of his troubles upon himself. In the summer of 1934, for example, Stark had arranged a public declaration of support for Adolf Hitler that Heisenberg had refused to sign. The excuses he gave for his past conduct were no defense. Nevertheless, the official from the Saxon ministry assured the rector in Leipzig that Heisenberg would not be disciplined for this previous politically unacceptable conduct. Heisenberg would just have to have a little more patience and wait for Reich Ministry of Education to act.[472]

✠   ✠   ✠

Probation The last foreign lecture tour Heisenberg undertook before the coming war cast its shadow over international scientific relations was a trip to Holland in January 1939. The physical colloquium of the University of Leyden invited Heisenberg to give a talk on "the penetrating components of cosmic rays."[473] The trip was approved without any objection. As usual, Heisenberg was required to submit a report upon his return.[474] Heisenberg arrived in Leyden on 25 January 1939, and stayed with his colleague and friend Hendrik Antony Kramers, professor at the University of Leyden. Heisenberg gave his talk that afternoon before an audience that included physicists from the University of Amsterdam and the Philips Factory in Eindhoven. A long discussion followed in which Kramers, Hendrik Casimir, and other Dutch scientists took part. The colloquium continued the following day with presentations from Kramers' students and colleagues on pressing problems of modern physics.

Heisenberg also gave a lecture on nuclear forces at the Philips Company, which included a hundred researchers from Philips' scientific staff. After the talk, Heisenberg toured the impressive experimental apparatus in the company laboratory. On 28 January, Heisenberg went with Kramers to the Hague, and there, in cooperation with the German embassy in Holland, the two physicists visited Prince Bernhard zu Lippe. In the afternoon, Heisenberg heard a talk in Amsterdam on the magnetic properties of solid state materials. He then visited his colleague Jacob Clay to discuss cosmic radiation and returned to Germany that evening.[475]

In April 1939 Heisenberg proposed another trip. He wanted to participate in three prestigious and very visible international physics meetings: a June conference at the University of Chicago on cosmic radiation, a September meeting on nuclear physics at the Technical University of Zurich, and the October Solvay Conference in Brussels on the properties of elementary particles. His travel costs would be paid for by the organizers of the conferences, and Heisenberg wanted to stay in America for six weeks in order to visit several institutes.[476] The rector passed on the request together with the approval of the head of the Leipzig University Teachers League.[477] REM approved the trips without special comment.[478] However, neither of the last two conferences took place.

⚔   ⚔   ⚔

The SS Report on Heisenberg A day before Heisenberg's trips were approved, bureaucrats from another part of the National Socialist state completed a document that would silence *Deutsche Physik*,[479] rehabilitate modern theoretical physics, and change Heisenberg's life. The SS had finally finished with its thorough examination of Heisenberg and his work. The SS sent the report to the Party Chancellery. When the SS forwarded a copy to REM, it told the ministry that Heisenberg should be given another appointment, where this new professorship should be, and why this post was suitable. The SS report, which apparently forestalled a parallel investigation in the Party Chancellery, was definitive.[480]

Heisenberg could not be called to Munich, for that would be seen as a victory over the Party officials there. Members of Himmler's staff independently informed Heisenberg why he could not receive the Munich professorship. It was the vacant professorship for theoretical physics at the University of Vienna that the SS wanted to be Heisenberg's new home. Most of the physics professors in Vienna had joined the NSDAP when it was still illegal in Austria, and were politically and ideologically reliable. The SS was cautiously optimistic that this circle of physicists would awaken Heisenberg's interest in political events and eventually attract him to National Socialism.[481]

According to the SS, Heisenberg was a man of surpassing scientific reputation. His strength lay in the school of physicists he had trained, which included Siegfried Flügge and Carl Friedrich von Weizsäcker. As for the controversy raging over the foundations of physics, Heisenberg argued that no conflict was possible between experimental and theoretical physics, because every theoretical physicist regarded experimental physics as an absolute necessity for his own work. Moreover, the converse was also true.

Heisenberg preferred to make a sharp distinction between "good" and "bad" scientists and was willing to agree that physicists who were "divorced from true experience" (a vague classification used by advocates of *Deutsche Physik*) were poor. The SS argued that Heisenberg's concept of bad physicist could be regarded as equivalent to the concept of "non-Aryan" (*artfremde*) thinker in physics. In particular, Heisenberg had agreed that some of the Jewish physicists and "Aryan" physicists from Jewish schools of physics, for their "Jewish" physics, who had been attacked by Lenard and Stark, were bad physicists.

The SS admitted that Heisenberg had been trained in a school of "Jewish physics." Consequently, his first great successes like quantum mechanics were influenced by "non-Aryan" physics. However, according to the SS, Heisenberg's work had recently become more and more "Aryan" (*artgemässe*). For Heisenberg, the theory was merely the working hypothesis with which the experimenter investigates nature by means of suitable experiments. The-

Werner Heisenberg (middle) in military training, ca. 1937. (Courtesy of the Library and Archives of the Max Planck Society.)

ory confirmed by experiment was therefore the clear description of observations made in nature aided by the exact tools of mathematics.

The SS also gave Heisenberg good marks for character. He was a typical apolitical scholar but nevertheless ready at any time unconditionally to serve Germany, because, as he told the SS, "someone is either born as a good German or not." Furthermore, Heisenberg had a strong military record. As a teenager in Munich he had fought with the Lützow paramilitary force (*Freikorps*) against leftists during the revolution and short-lived Bavarian Soviet Republic following World War I.[482] After Germany repudiated the Treaty of Versailles in 1935 and announced that it would rearm, Heisenberg had volunteered for the Army reserve. Finally, during the crisis of September 1938, when war with Czechoslovakia was forestalled only by the infamous Munich conference,

Werner Heisenberg (far right) in military training, ca. 1937. (Courtesy of the Library and Archives of the Max Planck Society.)

where France and Britain forced Czechoslovakia to give up the Sudetenland to Germany, Heisenberg had volunteered to fight and was one of the many German soldiers standing on the front waiting to attack.

The SS added that unfortunately Heisenberg's political attitude had not been as clear as would have been desirable. He had declined to take part in an election rally in 1933 (one of the many elections manipulated by the National Socialists) because his foreign colleagues, with whom he had very good relations, might have misunderstood. When invited to sign Stark's declaration for Hitler, Heisenberg had declined. But the SS argued that in the mean time Heisenberg had become more and more convinced by the successes of National Socialism and was now positively inclined toward it. However, he still believed that, aside from the occasional participation in an instructional (i.e., indoctrination) camp or the like, an active political role for a university professor was inappropriate.

Finally, the SS hoped that Heisenberg could be brought to accept anti-Semitism. The report claimed that even Heisenberg now rejected the "excessive alienation by Jews of German living space."[483] A few weeks later Himmler informed Heisenberg personally that he would be called to Vienna and, exactly as Prandtl had requested, be allowed to publish his views in the *Zeitschrift für die gesamte Naturwissenschaft*, the house journal of *Deutsche Physik*.[484]

But the Ministry of Education could not send Heisenberg anywhere without the explicit permission of the Party Chancellery, which had veto power over all important appointments in Germany, including university professorships. The SS could merely provide an assessment of Heisenberg's character and suitability and make a suggestion. When shortly before Christmas the SS proposed sending Heisenberg to Vienna,[485] the Chancellery rejected it. Party officials responded that Heisenberg's political conduct, especially after the National Socialist seizure of power, made this call unacceptable.[486]

This conflict over the fate of Heisenberg was typical of the polycratic institutional rivalry under National Socialism. Different agencies jealously guarded their own authority and sought to usurp that of others. No one power bloc, not even a force as powerful as the SS, could consistently dominate the others and get its way. In June 1939 the Party Chancellery learned that Heisenberg's three foreign trips had been sanctioned—which suggests that some REM officials opposed such permission—and pointedly reminded the Education Ministry that the Party had already opposed two proposed appointments for Heisenberg because of his political conduct. Conceding that it was too late to do anything about the trip to the U.S.A., the Party officials wanted the opportunity to express an opinion with respect to the Zurich and Brussels conferences, that is, to reverse the decision made by the ministry.[487]

But the Ministry of Education, now supported by the SS report on Heisenberg, stood its ground. Abraham Esau, a Party member since the spring of 1933[488] and a physicist with considerable political and professional influence, was to lead the massive German delegation to Zurich.[489] He intervened on Heisenberg's behalf. Esau had often had the opportunity to observe Heisenberg at international meetings, where, he said, Heisenberg had always conducted himself in a completely unobjectionable manner. Moreover, with respect to the prestige of German science, Esau emphasized that Heisenberg's presence in Zurich was very desirable.[490]

REM pointed out to the Party Chancellery that the local leader of the University Teachers League, the responsible Party official, had no political objection, and that Heisenberg was going to be one of the major speakers at the Zurich and Brussels meetings. Although in the past the Party had successfully put pressure on the Ministry of Education, this time Minister Rust politely told his colleagues in the Party Chancellery that they would have to live with his decision.[491] Heisenberg was too hot to be rewarded with a prestigious professorship, but he could be used as a propaganda tool.

✠   ✠   ✠

Lightning War and New Opportunities For Cultural Propaganda The German invasion of Poland in September 1939 represented a turning point for Heisenberg the itinerant lecturer. Whereas he had previously represented German science at international conferences, now he became a goodwill ambassador for the German war effort and, whether he liked it or not, for National Socialism. A reserve officer, Heisenberg was called up in September 1939,[492] conscripted by Army Ordnance for military research on nuclear fission, and allowed to return to his teaching in Leipzig a week later.[493] Heisenberg hoped that the conflict would not cost too many lives—unfortunately, he was wrong.[494] Most Germans were unenthusiastic about the war when it began.[495] Heisenberg was no exception, yet he was also determined to help his fatherland win the war.

The successful Lightning War provided new opportunities for National Socialist cultural policy outside of the Reich. Germany attacked, defeated, and occupied most of Europe in quick succession: Poland, Denmark and Norway, Holland, Belgium, Luxembourg, and finally France. Henceforth the great majority of Heisenberg's guest lectures would take place in countries either occupied by or obedient to Germany. Each trip required extensive approvals and notifications: the cultural-political section of the Foreign Office, the foreign branch of the NSDAP, the German Congress Center, and the German Academic Exchange Service all had a say. Most important, in the country to be visited the "German Cultural Institute" (GCI), which was under jurisdiction of the Foreign Office, or the local branch of the Exchange Service was to be informed.

The traveler had to acquire the necessary exit visa, foreign currency, leave from military service, and tickets. Foreign currency could be requested from the Congress Center only after REM had approved the trip. The Congress Center was to be informed of the exact duration, travel schedule, and any intermediate stops for the trip, as well as the exact topic of the lecture. Once the scholar had entered the foreign country, he had to immediately contact the official German delegation and either the GCI or Exchange Service.

GCIs, branches of the Exchange Service, or comparable institutions existed in Belgium, Bulgaria, Croatia, Denmark, France, Greece, Italy, Holland, Hungary, Norway, Portugal, Rumania, Serbia, Slovakia, Spain, and Sweden. In France and Belgium the traveler was to visit the military occupation authorities, in Norway the Reich Commissioner for the Occupied Norwegian Territories. If at all possible, the scholar was ordered to drop in on the foreign branch of the NSDAP. Once in the foreign country, if a scientist was asked to give an additional talk, then he had to ask permission from the German embassy. He also had to submit a report to REM upon his return, including discussions of his general impressions and experiences, his contacts with foreign colleagues, and the local attitude toward Germany and German policy.[496]

Special rules applied to different countries. Scholars in the protectorates of Bohemia and Moravia, parts of what had been Czechoslovakia, could attend conferences only in foreign countries as part of the German delegation, and if they wanted to speak a language other than Czechoslovakian, it had to be German.[497] Czech scientists could not lecture in Germany; indeed the German occupying authorities made few exceptions to their policy of not allowing any foreign scholars to travel to Germany.[498] Lecturers visiting Hungary and Rumania, both allies of Germany, were forbidden to discuss the relations between the two countries, especially their border dispute.[499]

Trips by German scholars to the General Government, part of what had been Poland, were placed under especially stringent restrictions. Any and all contact between German scientists and Polish colleagues was forbidden.[500] The General Government was in a sense a laboratory for the most extreme National Socialist policies, including German colonialism, slave labor, and from 1941 onward, genocide.[501]

The German Foreign Office and Ministry of Education together worked out guidelines for German scholars suitable to represent Germany in neutral (and presumably occupied or puppet) countries during the war. The scholar not only had to be a good scientist, he had to be well-known outside of Germany and able to contact his foreign colleagues immediately. Furthermore, the scientist had to show complete understanding of National Socialist domestic and foreign policy. Being apolitical did not suffice. Finally, the scientist had to possess social graces and, where necessary, knowledge of foreign languages.[502] The German authorities continued to use Heisenberg as a guest speaker, but since he stubbornly maintained his apolitical nature, the responsible officials became more and more ambivalent about his value for cultural propaganda.

As the program grew, officials became concerned about the uneven quality of the lectures by its touring scholars. Several reports of poor performances provoked threats and new guidelines from the Education Ministry. The speaker had to make a clear decision whether he intended his talk for a general audience or for a group of specialists. Every lecture was to be seen as a scientific performance and as a contribution to the cultural and political status of Germany. A lecture before academics which merely repeated known results and offered nothing new harmed the prestige of Germany as well as the personal reputation of the scholar.

Scientists who spoke to general audiences should also speak to a closed circle, seminar, or institute in order to make contacts with the foreign experts in their field. Finally, lecture topics should be chosen so as to offer something new to scholars outside of Germany. The Ministry gave the deans and rectors responsibility to judge the quality of the scientist when approving their applications to speak abroad. If valid criticism was made of a speaker, then REM would not allow him to travel abroad again.[503]

Since the speaker usually knew little about the political situation in the country he visited, the ministry suggested that he discuss the text of the lecture beforehand either with the GCI or the cultural department of the German mission. In September 1942 the SS informed the ministry that severe restrictions were being placed on any and all written materials taken across German borders. Any document, including the text of a lecture, had to be submitted beforehand for inspection and approval by the university intelligence officer.[504] In principle, the German scholar was instructed to avoid politically controversial topics while abroad. The scientist lectured in order to impress the natives with German culture, taking pains not to cause problems for the German political authorities or representatives.[505]

In November 1940, Heisenberg received an invitation through the German Foreign Office to speak at the Paris "German Institute" on "The current goals of physical research." Around the same time, Heisenberg was asked by the Hungarian "Union for Cultural Cooperation" to come to Budapest in early 1941 to deliver a paper on "Newton's and Goethe's theory of colors in the light of modern physics." Since Heisenberg was technically considered a soldier, he assumed that only the Army had to approve his talks, and that he did not have to consult REM.[506] The University told him that he was mistaken.[507]

Heisenberg dutifully wrote the ministry, noting that he had a letter from his superior in the Army granting him permission to give the talks.[508] The Leipzig representative of the University Teachers League supported the request, noting that Heisenberg was suitable in every respect to represent German science in foreign countries.[509] Both the dean and the rector agreed that Heisenberg was an appropriate candidate as well.[510] REM responded by rejecting the Paris trip[511] and approving the lecture in Budapest.[512] Apparently the distinction between a conquered enemy and an ally was important.

In May 1941, Heisenberg received an invitation to speak at the "German Institute for Eastern Work," located in the General Government.[513] The Germans had set up the institute at the site of the former University of Krakow. With very few exceptions, the Polish faculty of this university had been arrested by the German occupation forces and had been sent to the concentration camp in Sachsenhausen. Hans Frank, the governor of what in effect was a German colony on the eastern border of the Reich, was also the founder and promoter of this institute. The Institute's goal was to prepare for German expansion into this region by providing preliminary scientific research for German colonization of eastern Europe.

The Institute's work anticipated future "eastern research" of the sort that the National Socialists needed for their policy of acquiring "living space" for Germans at the expense of other peoples. For example, the Institute's section for astronomy and mathematics employed the forced labor of Russian prisoners of war and concentration camp inmates for mathematical research.[514] Wilhelm Coblitz, institute director, stated in 1941 that the Eastern Jewish question required scientific investigation as preparation for the final postwar solution of the European Jewish question.[515]

The invitation to speak in Krakow had originated with the governor himself.[516] Frank had been a schoolmate of Heisenberg's and may well have wished to show off one of the scientific institutes under his control. Heisenberg was willing to go.[517] The rector in Leipzig thought that he was perfectly suited for a foreign trip, both in the scientific and social senses.[518] A month later the officials in Leipzig sent on an additional letter from the Army, granting

Heisenberg permission to travel to the General Government.[519] But in 1941 when Coblitz asked REM for permission for Heisenberg to hold a lecture at the German Institute for Eastern Work, the request was denied.[520]

The German Institute for Eastern Work did not give up easily. Coblitz pointed out it was the personal wish of Governor Frank that Heisenberg be invited to Krakow. The ministry did not give permission, but provided an explanation. Heisenberg was a politically controversial figure. Because his connections to Jewish physicists and their followers in foreign countries were so extensive, the Party Chancellery had rejected two attempts to call this talented scholar to universities in Munich and Vienna.

Moreover, the Education Ministry understood the concerns of the Party. The Ministry of Propaganda had monitored Heisenberg's talk in Budapest and judged it unacceptable from the standpoint of National Socialism. All of his foreign talks were apolitical popular or specialized scientific lectures. The main problem in Hungary was his audience. The local "Jewish-influenced" physics community attended and enthusiastically applauded Heisenberg's lecture—no doubt embarrassing the National Socialist officials who were also present. Heisenberg could not go to Krakow, but REM assured Frank that it was more than willing to assist his cultural policy in any way it could. Frank had only to ask.[521]

☩   ☩   ☩

The German Astrophysics Conference at the Copenhagen German Cultural Institute In March 1941 Heisenberg's friend, colleague, and former student Carl Friedrich von Weizsäcker held several lectures in occupied Copenhagen and thereby set into motion a series of policy decisions that led to Heisenberg's most controversial foreign lecture. Von Weizsäcker spoke before the Danish Physical and Astronomical Society on "Is the world infinite in time and space?" The lecture, given in Danish, was both well attended and successful. He repeated the performance at the collaborationist Danish-German Society.

The German occupation authorities reported that von Weizsäcker knew how to make a difficult topic stimulating. The lay audience, including the commander of the German troops in Denmark, could follow it without difficulty. Finally von Weizsäcker took up an invitation from Niels Bohr's Institute of Theoretical Physics, and before a purely scientific audience, spoke on "The relationship between quantum mechanics and Kantian philosophy." A lively discussion followed. Although von Weizsäcker's conclusions were controversial, he managed to convince many of his Danish colleagues. Clearly the occupation authorities were also well informed about what went on in Bohr's institute.

The official report on von Weizsäcker's talks in Denmark judged that he had an exceptionally good influence on both lay audiences and purely scientific Danish circles. The German authorities in Denmark wanted to invite von Weizsäcker back to Copenhagen in the fall, this time together with Heisenberg, as part of a week-long conference on mathematics, astronomy, and theoretical physics at the newly-founded German Cultural Institute (GCI).[522] The German Foreign Office forwarded the request to REM with its approval.[523]

The initiative for Heisenberg's invitation came from von Weiszäcker, who has recently recalled that their concern about their Danish mentor Niels Bohr was one of the main reasons for their desire to visit Copenhagen.[524] Since Bohr's mother was Jewish, the German occupation officials considered him a "non-Aryan." However, Bohr and the other scientists at his institute had been able to continue work because during the first few years of the war Germany treated both Denmark and the Danish Jews relatively gently as part of the fiction that the Danish government had invited the German forces and was cooperating with the Third Reich.

A month later REM agreed that von Weiszäcker should return to Copenhagen, but ignored Heisenberg. The Kaiser Wilhelm Society, von Weiszäcker's employer, told the Minister that von Weiszäcker would be happy to take part in the Copenhagen conference.[525] The Ministry of Education, in turn, informed the Foreign Office in early June that von Weizsäcker would come.[526] But the German Cultural Institute wanted Heisenberg too.[527]

On 14 July von Weizsäcker met with an official from the German Academic Exchange Service in order to plan the Copenhagen conference. A week later he submitted a written proposal. Three German astronomers, Hans Kienle, Albrecht Unsöld, and Ludwig Biermann, should be invited along with von Weizsäcker and Heisenberg. The common theme of the conference could be the composition of the atmospheres of stars, a subject for which Kienle represented the best German empirical work, Unsöld and Biermann the best theoretical. In addition—and probably the main reason for the choice—the subject was also the main field of research for the Danish director of the Copenhagen observatory, Bengt Strømgren. Heisenberg would present his own work on cosmic radiation, while von Weizsäcker would discuss the transformation of elements in stars.

In his letter, von Weizsäcker recommended Heisenberg, as the leading theoretical physicist in Germany and someone who could not be surpassed for cultural propaganda. Since Heisenberg had spent years in Denmark and spoke fluent Danish, his participation in a conference in Copenhagen would be especially effective.[528] The Foreign Office informed REM in early August that both Heisenberg and von Weizsäcker had been consulted, and asked whether the authorities in Copenhagen could count on the participation of Kienle, Biermann, and Unsöld as well.[529] Von Weizsäcker wrote to Bohr, informed him that he and Heisenberg were going to speak at the astrophysics conference at the GCI, and invited all of the Danish scientists to attend.[530]

But REM, which had just turned down Frank, resisted the idea of sending Heisenberg to Copenhagen. They argued that a conference in astronomy had already been planned for Würzburg for October 1941, that many foreigners and especially Danes had been invited, and that the special event desired by von Weizsäcker overlapped with, and would detract from, the Würzburg meeting. Additionally, the ministerial official criticized the choice of scientists proposed for the Copenhagen meeting. Heinrich Vogt, Heinrich Siedentopf, Bruno Thüring, and Paul ten Bruggencate—all politically acceptable to the National Socialist state—were supposedly the leading German scientists in the field of the atmospheres of stars.[531] The ministry wanted to use the Würzburg meeting to abort the Copenhagen conference. A decree to this effect was drafted, but never sent.[532] The Foreign Office intervened again and requested a meeting with REM.[533]

An official from the Foreign Office, the director of the Copenhagen GCI, and a representative of the ministry got together on 2 September. The director pointed out that the conference had already been announced. A cancellation now, when the GCI was just beginning its work in Copenhagen, would be very damaging. The objections voiced by REM were irrelevant. The GCI did not particularly care what the theme of the conference was, or—with an obvious exception—which Germans took part. The meeting in Copenhagen would be a scientific colloquium and have no official character. The Würzburg conference would not be harmed, especially since the two Strømgrens—father and son—were going to Würzburg as well. Moreover, Heisenberg would only be in Copenhagen for two or three days.[534]

After some discussion, a proposal was cleared with Rudolf Mentzel, the head of the science section in the ministry,[535] to pass the buck. The Education Ministry would approve the conference if the Party Chancellery approved Heisenberg's participation. The head of the Cultural Political Section of the Foreign Office considered the matter very important. If the Copenhagen conference was rejected, then State Secretary Ernst von Weiszäcker, the father of Carl Friedrich, would intervene. Thus for tactical reasons it was desirable that von Weiszäcker's proposal be approved.[536]

The Education Ministry accordingly wrote to the Party Chancellery that von Weiszäcker, in close cooperation with the GCI in Copenhagen and after successful lectures in Denmark, wished to hold the proposed conference in Copenhagen, at which Danish and German scientists, including Heisenberg, were to take part. The workshop would take place in the GCI without being advertised to the greater public. Did the Party object to Heisenberg's attendance? Given the need for haste, the ministry telephoned the Party Chancellery in order to hear the decision as soon as possible.[537]

The Party Chancellery responded that there was no objection to Heisenberg's going to Copenhagen, provided that he kept a low profile and stayed only a few days.[538] This decision went out the day before the rejection of Heisenberg's trip to Krakow.[539] The Foreign Office was more powerful than Frank, and Denmark a less sensitive area than the General Government. The Party did take care to emphasize once again that a high profile visit from Heisenberg was undesirable.[540]

Heisenberg, von Weiszäcker, the German occupation authorities, and later, the Danish scientists, all wrote reports of this visit. Heisenberg evaluated opportunities for Danish-German cultural relations poorly. Because he had to return to Germany before the conference was over for personal reasons, Heisenberg received permission from the Foreign Office to go to Copenhagen a few days early. He was welcomed by an official from the GCI on 15 September, met with Strømgren at the Copenhagen Observatory the following day, when he agreed on the schedule for the workshop, and contacted his colleagues at Bohr's institute.

The meeting began on 19 September. The only Danes who attended were the two Strømgrens and the staff of the observatory. The physicists from Bohr's institute boycotted the conference. Several members of the German colony in Copenhagen appeared just in time for Heisenberg's talk on cosmic radiation. Afterward, Heisenberg met with the NSDAP representative in Denmark and the following afternoon the German scientists were the guests of the German ambassador in Copenhagen. On 21 September Heisenberg left Denmark.

German relations with scientific circles in Scandinavia had become very difficult, he wrote in his report. Everywhere he went, he encountered a very reserved, if not dismissive attitude. Very few Danish colleagues were prepared to engage in scientific cooperation within an official institution like the GCI. Heisenberg concluded with a nonsequitur. The Danes took this position even though almost all of his Danish colleagues told him that they did not have the slightest criticism of the conduct of German troops in Denmark. Where Heisenberg's Danish colleagues saw "Nazi" invaders, he saw German soldiers.[541]

Von Weiszäcker tried to present a positive picture. Instead of mentioning that most Danish scientists boycotted the meeting, he emphasized that five did attend, and that the meeting was exceptionally fruitful. Instead of referring to members of the German Colony in the audience, von Weiszäcker noted that representatives of the German occupation government and the NSDAP attended, as well as at least one other Dane, the rector of the University of Copenhagen. Von Weiszäcker argued that the conference was living proof that scientific research continued in Germany despite the war, and ended rather weakly by suggesting that the opportunity in personal conversations to set right several false judgments about Germany was "not without significance."[542]

At the end of the war, Danish scientists explained that they perceived the policy of the GCI as an attempt to coerce Bohr and his colleagues into cultural collaboration. Although pressed to attend the lectures—von Weiszäcker told the Danes that if they did not come to the GCI, then the SS would open their own cultural institute—the Danes refused. During the conference, von Weiszäcker brought the director of the GCI into the Institute of Theoretical Physics and pushed him without an appointment past Bohr's secretary. Von Weiszäcker thereby forced Bohr into a confrontation he had taken pains to avoid, in part because he feared that the Danish resistance would believe that he was collaborating with the Germans. The Danish scientists also recalled that Heisenberg had callously offended them by remarking that war was a "biological necessity" and behaving as an intense nationalist, with the characteristic German deference to authority, here to the German state.[543]

In 1961, Bohr told a Soviet colleague a similar story. Heisenberg came to Bohr in the autumn of 1941, when Hitler had already defeated France and was advancing quickly into Russia. Heisenberg had wanted to convince his mentor that Hitler's victory was inevitable and that it would be unwise to doubt it. The National Socialists did not honor science, which was why they treated scientists so badly. Bohr had to join forces with Heisenberg and help Hitler. When the National Socialists were victorious, then their attitude towards scientists would change. In particular, Heisenberg told Bohr that he had to cooperate with the GCI.[544]

Moreover, Heisenberg made similar statements after the war. In their obituary for Heisenberg, Neville Mott and Rudolf Peierls gently criticized him for his obtuseness. When Heisenberg visited a German refugee physicist in Great Britain late in 1947, Heisenberg argued that if the National Socialists had been left in power for another fifty years, then they would have become quite decent. As Mott and Peierls note, that was a strange remark to make to a colleague who had first lost his job and then relatives and friends in extermination camps.[545]

Perhaps most interesting, the report of the 1941 visit from the German authorities in Copenhagen was very positive. According to an official from the German occupation forces, the workshop had been run by the Danish scientist Strømgren and the significant Danish astronomers as well as some theoretical physicists had attended. This German official was also the only reporter who mentioned that the German physicists Walther Bothe and Kurt Diebner, both of whom were involved with the Army research into the military applications of nuclear fission, participated in the conference as well. In the opinion of the German officials in Copenhagen, both the workshop and the popular lectures at the GCI were great successes, for they drew new Danish researchers into the GCI.[546] That had been the purpose all along.

The Foreign Office did not stop there. In November 1941, it informed the Ministry of Education that the Party Chancellery intended to make a definitive decision: should Heisenberg be used for foreign lectures in the future? The Foreign Office had no doubt that with regard to cultural political considerations, Heisenberg was extremely valuable. The reports on his lectures in foreign countries—and here the report on Budapest seems conveniently to have been forgotten—had all been very positive. Moreover, several independent suggestions had been made for using Heisenberg more often as a guest lecturer. The Foreign Office wanted to know: was Heisenberg an acceptable goodwill ambassador for German culture or not?[547]

There is one important aspect of Heisenberg's and von Weiszäcker's 1941 visit with Niels Bohr which Heisenberg and von Weiszäcker rarely mentioned in their many postwar descriptions of the event. When they traveled to Copenhagen, the German Lightning War was driving deep into the Soviet Union. Most Germans, and most probably Heisenberg and von Weiszäcker, believed that Hitler's victory was imminent. It is unlikely that the two German physicists would have been concerned about the prospect of developing nuclear weapons for this war.

The historian Philippe Burrin has convincingly argued that the decision to launch the Holocaust, the physical extermination of all Jews under German control, was made on 18 September 1941, one day before the conference began at the Copenhagen German Cultural Institute.[548] Of course it took some time before the National Socialist leadership's policy change, from forcing the Jews to emigrate or planning to concentrate them on a "reservation" to murdering them, would become known to Germans like Heisenberg or conquered nationals like Bohr. But in retrospect, the German astrophysics conference in September 1941 was a watershed in many respects. Up until this point, Heisenberg had consciously or unconsciously been a goodwill ambassador for National Socialism and German military aggression. Henceforth he would consciously or unconsciously be an ambassador for genocide.

7

Goodwill Ambassadors

Rehabilitation Ludwig Prandtl made a second, more vigorous assault on National Socialist policy towards physics in the spring of 1941, this time seeking allies in German industry, including Carl Ramsauer, a leading physicist at German General Electric.[549] Germany's misfortune in war also played into the hands of Prandtl, Ramsauer, and company. Shortly after the Soviet defense had frozen the Lightning War in its tracks during the winter of 1941, it was clear that the entire German war economy had to be reorganized and made more efficient. Although victory still appeared possible, the war now appeared much more difficult to win.

Ramsauer now succeeded in convincing Major General Friedrich Fromm, the commander of the German Reserve Army and chief of armaments production, that German physics, and with it Germany's ability to wage war, was in grave danger.[550] By early December 1941, Prandtl had received a favorable response from Field Marshall Erhard Milch, Hermann Göring's deputy in the Air Force Ministry.[551] The Air Force appreciated the connection between academic physics and the industrial production of modern weapons.[552] After assembling such powerful political backing, Ramsauer submitted a twenty-eight page memorandum with six appendices on the sorry state of German physics to REM.[553] Ramsauer did not expect Rust to react to this challenge, nor did he, but the Ministry of Education was not the main target.[554] Ramsauer's memorandum circulated widely. The highest agencies of the government, including the military, developed a great interest in theoretical physics.[555]

Perhaps the best example of such interest was the popular nuclear fission lecture series held on 26 February 1942 in Berlin-Dahlem before a restricted audience of representatives of the National Socialist Party, the German state, and German industry.[556] Minister of Education Bernhard Rust, Albert Vögler, the President of both the Kaiser Wilhelm Society and Germany's largest steel concern, and the Reich Research Council were in attendance.[557] Along with popular talks on the latest research results given by the responsible project scientists, the Army representative Erich Schumann discussed the military applications of nuclear fission, the Reich Research Council representative Abraham Esau stressed the significance of nuclear power for the state and industry, and Hans Geiger, a politically and professionally very conservative experimental physicist, made the connection between research and application.[558]

These lectures gave the members of the nuclear power project the opportunity to sell their research for financial, material, and institutional support. The vivid and suggestive contributions by Otto Hahn,[559] Paul Harteck,[560] and Heisenberg[561] were exemplary in this respect. Hahn did not mention his former Jewish collaborator Lise Meitner in his historical account of the discovery of nuclear fission; instead he described enthusiastically the potential of nuclear-fission chain reactions.[562] Harteck was even more colorful in his justification of heavy water research. Heavy water could be used to ignite a nuclear fission chain reaction. Once lighted, no one knew how long or how powerfully this flame could burn.[563]

Heisenberg used a diagram of the various possible nuclear reactions in uranium and moderator to provide his listeners with

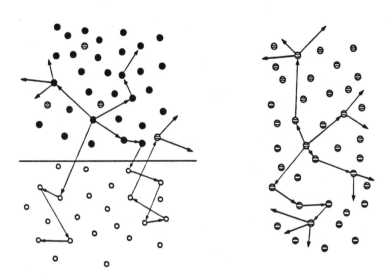

Chain reaction in uranium machines (left) and in nuclear explosives (right). The solid black circles represent uranium 238, the ruled circles uranium 235, and the small circles moderator. (From Walker, p. 56.)

a layman's description of how uranium machines and nuclear explosives should work (see diagram on page 155).[564] The left-hand portion of the diagram represented a schematic uranium machine and the various nuclear processes that a fission neutron could experience in uranium. A fast neutron can fission a uranium 238 nucleus, but, as Heisenberg realized, with very low probability. After a few collisions, the slowed neutron might be absorbed by a uranium 238 nucleus, and disappear from the scene. If, instead, the slow neutron collided with a uranium 235 nucleus, it might cause fission. But that was very unlikely. Therefore the desired chain reaction could not proceed in ordinary uranium; new techniques were needed in order to force the chain reaction.[565]

Heisenberg then made an analogy both in the spirit of the times and tailored to the level of comprehension of his audience.[566] The behavior of neutrons in uranium could be thought of as a human population, where the fission process represented an analogy to a child-bearing marriage and the neutron capture process corresponded to death. In ordinary uranium, the death count overwhelms the birth rate, so that a population must die out after a short period of time. For survival, the number of births per marriage or the number of marriages must be increased, or the probability of death reduced.

Heisenberg told his audience that nature prohibited an increase in neutron births. An increase in the number of fissions/marriages could be achieved by enriching the uranium 235 in the uranium sample. If pure uranium 235 could be produced, Heisenberg noted, then the processes represented in the right-hand side of the diagram could take place. Unless a fission neutron escapes through the outer surface of the uranium, every neutron would cause a further fission after one or two collisions. In this case, the probability of death was vanishingly small compared to the likelihood of neutron increase.

If a large enough amount of uranium 235 could be accumulated, then the number of neutrons in the uranium would increase tremendously in a very short period of time. The isotope uranium 235 might make an explosive of "utterly unimaginable effect." Heisenberg hastened to inform his audience of prospective patrons that the explosive uranium 235 was very difficult to obtain. As for reducing the probability of neutron death, Heisenberg noted that a uranium machine composed of uranium and a neutron moderator could facilitate fission in uranium 235 without great danger of neutron absorption by the heavier isotope uranium 238. Heisenberg observed that, like uranium 235, large amounts of the moderator heavy water were not easy to obtain.

Heisenberg recommended uranium machines as heat engines which could produce energy and power vehicles or ships. These machines would be particularly suitable for submarines, since a nuclear reactor does not consume oxygen. But these uranium machines had an even more important application. The transformation of uranium in the machine created a new substance, element 94 (plutonium), which most probably would be as explosive as uranium 235, and much easier to manufacture since it could be separated chemically from its parent. Uranium enrichment made nuclear energy and explosives possible. A uranium machine could function as a heat engine and produce another unimaginably powerful explosive. To achieve all this, Heisenberg recommended strong financial and institutional support for the nuclear power project. In short, Heisenberg went out of his way to illustrate clearly and vividly the warlike aspects of nuclear power.[567]

As Hahn noted in his diary, the lectures before the Reich Research Council made a good impression.[568] They were subsequently publicized in a newspaper account under the title, "Physics and National Defense." Although the words atomic, nuclear, energy, or power did not appear, a reader would have learned that the meeting dealt with problems of modern physics decisive for national defense and the entire German economy.[569] The physicist and Party official Wolfgang Finkelnburg could soon tell Heisenberg that his lecture before the Reich Research Council and the subsequent press accounts had had a good effect. Finkelnburg had received several inquiries from Party positions concerning the military importance of theoretical physics and especially of Heisenberg's work.[570]

The military potential of nuclear power penetrated into the highest circles of the National Socialist state. On 21 March, less than a month after Heisenberg's lecture, Reich Minister of Propaganda Josef Goebbels noted in his diary that he had received a report on the latest developments in German science. Goebbels learned that research on atomic weapons had progressed so far that it might be used in the ongoing war. His reports claimed that tremendous destruction could be wrought with a minimum of effort, with terrifying prospects for war. Modern technology placed means of destruction in the hands of human beings, the Reich Minister of Propaganda noted, that were incredible. It was essential that Germany be ahead of everybody, he recognized, for whoever could introduce such a revolutionary innovation into the war had the greater chance of winning it.[571]

By this time, no one involved with the research or administration of the nuclear power project believed that nuclear fission could influence the outcome of the war. But by dangling seductively the prospect of unimaginably powerful weapons sometime in the future, scientists from the German nuclear power project could, and did, enjoy exceptional political and financial support from several diverse sections of the National Socialist German state.

For example, in the spring of 1943 Hahn and Heisenberg lectured at the Reich Postal Ministry before a small circle of around fifteen people, including Postal Minister Ohnesorge, Minister of Armaments Speer, and General Keitel, head of the supreme command of the Armed Forces. Hans Meckel, a former staff member of the Navy commander Admiral Dönitz, attended this meeting and remembered one statement from Heisenberg very clearly: even though there were a few still unsolved problems, within one to two years the scientists hoped to be able to offer the National Socialist leadership a bomb with "hitherto unknown explosive and destructive power."[572]

The rehabilitation of modern physics and the great interest in nuclear power improved Heisenberg's position in the National Socialist state. In June 1942, he became director of the Kaiser Wilhelm Institute for Physics in Berlin-Dahlem. A professorship at the University of Berlin usually went with the directorship. The planned appointment caused another round of political reports on Heisenberg from various branches of the NSDAP. These investigations[573] cleared the way for Heisenberg's call to Berlin. The unlikely combination of the SS's positive report and the newly found support for modern physics in German industry had fully rehabilitated him.

The Ministry of Education stressed the importance of Heisenberg's appointment for the national defense. Both Albert Speer's Ministry of Armaments and the Armed Forces had great interest in Heisenberg's research.[574] Indeed Heisenberg subsequently told a colleague that Speer took a great personal interest in nuclear physics research.[575] Alfred Rosenberg's office echoed Ramsauer's memorandum and argued that the Party could not intervene in the "difference of opinion" between Lenard's and Heisenberg's schools of physics.[576] The Reich University Teachers League merely repeated some of the positive statements made about Heisenberg in the SS report and added pointedly that Himmler had personally called a halt to political attacks on Heisenberg.[577] The contrast with the previous attempts to bring him to Munich and Vienna is stark.

⌘   ⌘   ⌘

Lectures in Switzerland and Budapest In the spring of 1942, Heisenberg received an invitation to speak before the Swiss League of Students. Switzerland was one of the few countries in Europe to remain neutral during the war. The Swiss physicist Paul Scherrer, who had recommended his German colleague for the lecture, asked Heisenberg to give a talk before the physicists at the Zurich Technical University as well.[578] Heisenberg became inundated with offers for speaking engagements. In the end, he agreed to lecture before the Science Faculty of the University of Geneva, the Swiss Physical Society, and the student organizations of Bern and Basle as well.[579] The rector at the University of Leipzig noted as usual that the dean considered Heisenberg suitable for the trip and that the University Teachers League representative had no objections. He asked REM for its approval,[580] which was granted in late October.[581] The Party reminded him of his obligation to call upon its foreign branch while in Switzerland.[582]

On 17 November 1942, Heisenberg arrived in Zurich and was met by the head of the Swiss Students League. The next day, he spoke at the university colloquium on the observable variables in the theory of elementary particles. Afterward he visited his old colleague Scherrer at the Technical University. Heisenberg's next lecture came before the Swiss Physical Society on 19 November, which included dinner afterward as the guest of the president of this society. The next day he went to Basle, paid a courtesy call on the physicists there, and in the evening spoke before the local student organization on the current goals of physical research.

Two days later, he gave an evening lecture before the Zurich student organization on changes in the foundation of the exact sciences. On 24 November, he visited the German ambassador to Switzerland and the representative of the Party in Bern and lectured to the Bern student organization. Heisenberg reported that he was treated throughout in a very friendly fashion in Switzerland, and not just by old colleagues. He encountered frequent political condemnation of the German "re-ordering" of Europe, but this ill will did not carry over to personal relationships. His lectures had attracted great interest.[583]

In October 1942, the German ambassador to Hungary, a German ally, complained to the Foreign Office about REM's unwillingness to allow Heisenberg to return to Budapest. With his Nobel Prize and his call to the Kaiser Wilhelm Society, Heisenberg was so well known that a lecture from him guaranteed a cultural and political success. Hans Freyer, who had been professor for philosophy and sociology at Kiel and Leipzig during the Weimar Republic, and who was now the president of the Budapest GCI, wanted to invite Heisenberg for a talk in his institute. However, Freyer did agreed that, because of the controversy Heisenberg's previous trip to Hungary had caused, other lectures in Budapest would not be a good idea.[584]

The Budapest GCI managed to get around the recalcitrant ministry by joining forces with the Kaiser Wilhelm Society. In early November the Society informed the ministry that a joint scientific meeting had been planned with the Budapest GCI, including talks not only by Heisenberg, but also from Max Planck and Carl Friedrich von Weizsäcker.[585] The Education Ministry reacted angrily. Another talk by Heisenberg in Budapest would undoubtedly attract foreign scholars of Jewish origin or liberal political views who had been connected with German physics before the National Socialists took power. For example, Heisenberg had former students and colleagues in Hungary. The ministry was afraid that some members of the audience would see the affair as a political demonstration for Jewish scientists.

However, the request by the Budapest GCI was very much strengthened by the Kaiser Wilhelm Society's participation.[586] The joint series which would present Heisenberg along with Planck and von Weizsäcker to the Hungarian public would not be easy to cancel. REM informed the Foreign Office that they considered German initiatives for sending Heisenberg abroad inappropriate because his visits always ended up being so controversial. But since Ernst Telschow, General Secretary of the Kaiser Wilhelm Society, had gone so far ahead with preparations for the lectures without consulting either the ministry or the Foreign Office, REM agreed to go along—this time.[587]

Heisenberg, von Weizsäcker, Planck, and the German ambassador to Hungary submitted reports on the lectures. Heisenberg's was the most sober. On 30 November 1942 he arrived in Budapest and joined Planck and von Weizsäcker as the guests of the Budapest institute. Planck and von Weizsäcker spoke on the first two days of December, respectively. Heisenberg had lunch with the director of the GCI on 2 December, tea with the German ambassador to Hungary, and lectured that evening on "the current goals of physical research." An informal party at the institute brought the activities of the day to a congenial close.

The three German physicists met the physics professor at the University of Budapest for lunch on the following day and Heisenberg joined his counterpart at the local technical university for dinner. He returned to Germany on 4 December. When Heisenberg reported his impressions of the political climate in Budapest, he judged that the GCI had succeeded in keeping alive the Hungarian interest in German cultural goods in a most auspicious manner.[588]

Von Weizsäcker reported that he spoke on "atomic theory and philosophy" before invited guests, including officials and the representatives of physics and the neighboring disciplines at the local universities. After the talks, he had a pleasant opportunity to meet with Hungarian colleagues. Von Weizsäcker's remarks about the Budapest trip stand in sharp contrast to his 1941 report on the

Copenhagen conference. The apparent interest in cultural politics he showed at that time disappeared shortly after the tense meeting in Denmark, never to return.[589]

Planck's report enthusiastically praised the export of German culture. He gave his standard talk on "The senses and boundaries of the exact sciences." The president of the GCI, who as Planck noted approvingly had set himself the task of cultivating the cultural relations between Germany and Hungary, met Planck and his wife at the train station and looked after them throughout their stay. Planck's talk was held on 1 December in the cozy atmosphere of the GCI. Guests included representatives of the German delegation to Hungary of the NSDAP, and many Hungarian dignitaries, including Archduke Joseph, the president of the Hungarian Academy of Sciences, and the Archduchess Anna.

Following Planck's talk, an official reception was accompanied by pleasant personal conversation. Planck was impressed both by the good will towards Germans expressed by the Hungarians and especially by Freyer's exceptional skill. He understood how to awaken and maintain interest in German culture among the educated circles in Hungary. Planck reckoned that the entire event completely fulfilled its goal, to support the intellectual connections between Germany and Hungary.[590]

The account by the German foreign service stressed the collaboration of the Kaiser Wilhelm Society. The Budapest GCI previously had sponsored only lectures in the humanities; they decided to try physics in order to attract Hungarians interested in science. The Kaiser Wilhelm Society was happy to send a few scientists. At first its president, Albert Vögler, planned to attend as well and provide a brief survey of the society. General Secretary Ernst Telschow went instead. As Freyer noted approvingly, Telschow's talk provoked great interest among the Hungarian scientists and the representatives of the Ministry of Culture. The Hungarians had lost the research funds they previously had received from America; Freyer believed that Germany could fill the gap.

As far as the scientific talks were concerned, Freyer noted with approval that the aged Planck spoke with astonishing freshness, inner dignity, and intellectual elegance. Heisenberg, in his presentation of the current problems in physics and promising research areas, lectured with a clarity and maturity which only a researcher working on the furthest boundaries of science could provide. Von Weizsäcker, who spoke without notes, impressed Freyer with his ability to combine physics with philosophy so productively. The discussion provoked by von Weizsäcker's talk lasted until midnight. The lectures by Heisenberg and von Weizsäcker were followed by a concert of Bach and Mozart.

From the perspective of the president of the Budapest GCI, the lectures were a complete success. The audience had been hand-picked, and almost no invitations were declined. Along with the Archduke and Archduchess, the guests included the ambassadors or representatives of Italy, Finland, Croatia, and Slovakia, the Hungarian Minister of Culture, all the relevant professors from the University of Budapest, and representatives of other Hungarian universities. Best of all, great interest had already been expressed from the Hungarian side for more such cultural events, which was what Freyer wanted to hear.[591]

✠   ✠   ✠

The Goodwill Ambassador Heisenberg's trip to Budapest was the last time he experienced difficulty in traveling abroad. Henceforth, if he declined an invitation to speak, then it was his decision. The delayed effect of his dual appointment in Berlin and the ever-worsening state of the war inspired the change in policy.[592] Heisenberg's secret research had been classified important for the war. As the German position in the conflict deteriorated, his standing inside the National Socialist state climbed slowly but steadily, as demonstrated by his election to the Prussian Academy of Sciences in early 1943.[593]

Heisenberg received two invitations to France in 1943. The German Embassy in Paris was sponsoring a lecture series at the College de France, and wanted Heisenberg to deliver a strictly scientific talk in French.[594] The dean at the University of Berlin forwarded the invitation to Heisenberg with the remark that he, the rector, and the representative of the University Teachers League naturally would support the trip.[595] The German Institute in France also wanted a lecture from Heisenberg.[596] He turned down both offers because his French was not good enough for lecturing.[597] In contrast, Carl Friedrich von Weizsäcker did give a lecture in Paris, but both his talk and the lunch in his honor were boycotted by his French colleagues. When the French physicist Frédéric Joliot-Curie criticized his German colleague for the "bad taste" he had showed by accepting an invitation from the German occupation authorities, von Weizsäcker replied that he had been forced to accept.[598]

In February 1943, the Slovakian University in Pressburg (Bratislava) sent an invitation by way of REM for Heisenberg to lecture in the Slovakian Protectorate. In a striking about-face a ministry official now told Heisenberg that they wanted him to accept the invitation.[599] Heisenberg agreed to go.[600] On 28 March Heisenberg met the president of the local technical university, the dean of the Slovakian University, and a representative of the German Academic Exchange Service. That afternoon Heisenberg was the guest of the president, who took him to the opera in the evening. The next day, Heisenberg had an audience with the German ambassador, lunch with the dean and the president, an evening lecture on the state of atomic physics, and a late dinner with some Pressburg scientists. The following day brought more of the same: a walk through the old town hall with the mayor of Pressburg, lunch with the dean, the president, the local head of the German Academic Exchange Service, and the German ambassador, an evening lecture on cosmic radiation to a small group of scientists and students, and dinner with Pressburg scientists and a visiting Italian mathematician. The Pressburg scientists were very friendly. Heisenberg reported that the relations between Germans and their Slovakian colleagues were very good.[601]

A second popular lecture series on nuclear power was held before the Air Force Academy in May 1943.[602] By demonstrating the usefulness of modern physics, these lectures became part of the continuing battle against *Deutsche Physik*. Indeed Heisenberg's foreign lecture tours in general also contributed to the continuing campaign against the forces of Lenard and Stark—the advocates of *Deutsche Physik* were of no use when it came to foreign cultural propaganda. A month earlier, Carl Ramsauer had repeated his arguments about the dangerous decline of German physics before this same sympathetic audience. Since Ramsauer had kindled the interest of Academy members in nuclear physics, Heisenberg was asked to arrange a lecture series to keep it alive.[603]

Abraham Esau, the administrator in charge of nuclear physics research, opened the series with a status report on the nuclear power project and followed it with a talk on the production of luminous paints without the use of radium, a pressing topic for the manufacture of aircraft dials.[604] Otto Hahn spoke on the artificial transmutation of elements—and this time, before a less political audience, mentioned Lise Meitner by name as contributing to the work that led up to the discovery of nuclear fission.[605] Klaus Clusius discussed isotope separation,[606] and Walther Bothe lectured on the research tools of nuclear physics.[607] All of these speakers stressed the utility of physics as well as the need for increased governmental support.

Heisenberg's contribution paralleled his 1942 lecture before the Reich Research Council. But two differences are significant. In contrast to the winter of 1941–1942, the uranium research now enjoyed secure political and financial support; in contrast to Heisenberg's February 1942 talk, he now represented nuclear fission as irrelevant to the war effort. A chain reaction in uranium 235 would produce large amounts of energy explosively, Heisenberg noted, but that was as close as he came to mentioning nuclear explosives. He told his audience that the first step toward a very important technical development had been taken. Nuclear power could be liberated for large-scale applications. However, he closed on a more somber note. The practical execution of this process was greatly hindered by the strained economy and the great external difficulties presented by the war.[608]

Shortly before the lecture series before the Academy, Heisenberg received an invitation from the SS. In 1942 the first "SS-House" outside of the Reich had been established in Leyden. Himmler entrusted it with two tasks: providing Dutch students with a Germanic education and establishing contact with intellectuals in Holland. The Dutch were to become acquainted with German "ideological goods." In a year's time, the director of the SS-House believed that he and his colleagues had made a good start towards their cultural and political goals, but they recognized that the German military setbacks of the previous winter as well as political developments inside Holland had created difficulties. For this reason the SS decided to invite leading German scholars to Leyden in order to demonstrate the prowess of German intellectuals to Dutch academics. Heisenberg was asked to visit Leyden in the spring of 1943.[609] He declined because he was too busy, but encouraged another invitation in the fall. The SS apparently did not contact him again.[610]

In June 1943, the collaborationist Dutch Ministry of Education sent Heisenberg another invitation to visit Holland. The Reich Commissioner for the Occupied Dutch Territories, the highest German official in Holland, encouraged Heisenberg's acceptance.[611] REM welcomed the proposal, especially since the invitation had come from the Dutch Ministry.[612] Heisenberg told the ministry in Berlin that he was willing to visit Holland in principle, but only under certain conditions. He already had asked the Dutch officials to tell him which of his Dutch colleagues wanted to see him and what the exact details of his itinerary would be. He wanted to know what his Dutch colleagues—including friends and former students—thought of the idea before he committed himself.[613]

A Dutch official in the Dutch ministry collaborating with the German authorities called in Kramers and showed him Heisenberg's letter. Kramers wrote directly to Heisenberg to describe the poor working conditions of Dutch academics. An official of the Dutch Ministry of Education had intimated that this situation might be improved by reestablishing personal scientific contacts between Dutch and international—in other words, German—colleagues.

The Dutch and German authorities wanted Heisenberg to spend a week in Holland. He would visit all the physics institutes, meet with his Dutch colleagues, and give talks drawn from his own research before small groups of Dutch physicists. Thus Heisenberg's itinerary would fulfill the new governmental guidelines for foreign lectures. Kramers added that he had discussed this matter with Casimir and other Dutch scientists. All would welcome a visit by Heisenberg—which was exactly what Heisenberg wanted to hear.[614]

The adverse working conditions which Kramers mentioned may be illustrated by the state of the physical laboratory at the University of Leyden, where Kramers was professor of theoretical physics. German authorities had seized and closed the laboratory. The scientific equipment was to be shipped to Germany as war booty. Dutch scientists were prohibited from entering the laboratory.[615]

As soon as he received the letter from Kramers, Heisenberg told REM that he would visit Holland and implied that the personal invitation he had received from his Dutch colleague had been a crucial factor in his decision.[616] Heisenberg simultaneously wrote to Kramers and expressed his pleasure in the upcoming visit.[617] Kramers replied in kind.[618] The German officials were pleased that Heisenberg was coming, but also displeased that Kramers, who was not cooperating with the occupation authorities, had become involved.[619] They informed Heisenberg that although he was free to see Kramers informally, Kramers would not be an official participant in the program for Heisenberg's visit. Furthermore, Heisenberg was ordered to visit the German occupation authorities at the very beginning of his visit in order to be briefed on the political state of the Dutch universities.[620]

Heisenberg traveled to Holland in October 1943, following a summer of protests by students and professors at Dutch universities over German occupation policies, including the persecution of Dutch Jews. The Germans responded with harsh repression and deportations of Dutch Jews to the death camps.[621] As soon as Heisenberg arrived in the Netherlands, he met with collaborating officials from the Dutch Ministry of Education and with representatives of the German occupation authorities. The following day he paid a courtesy call on the physics institute in Utrecht, and dined with the theoretical physicist Leon Rosenfeld. In the morning Heisenberg journeyed to Leyden, visited the famous Kammerlingh-Onnes Laboratory, and met Kramers. On 21 October, Heisenberg gave the first talk of his trip, a lecture on the theory of elementary particles, at a small colloquium at the Leyden physics institute.

Heisenberg spent the next few days in Delft, where he visited his colleague Kronig as well as the nearby technical university. On 24 October, Heisenberg and the physicists from the Philips Company and from the University of Leyden attended an informal colloquium presented by Kramers at Rosenfeld's house. The next day Heisenberg traveled to Amsterdam, where the physicist participated in some experiments on cosmic radiation. On 26 October, Heisenberg discussed his visit with Dr. Seyss-Inquart, the German Commissioner in Holland. According to Heisenberg's subsequent report, everywhere he went he met a most cordial reception. He avoided politics wherever possible; when it did come up, Heisenberg reported, his Dutch colleagues harshly rejected the German point of view. However, he nevertheless assured his official readers that cooperation with the Dutch on a purely scientific basis was definitely possible.[622]

Shortly after the end of the war, Hendrik Casimir was questioned by the astronomer Gerard Kuiper, a former countryman and now a member of the American Armed Forces. Kuiper wrote a report that vividly captured the impression of callous nationalism that Heisenberg had made on his Dutch colleagues. According to Casimir, when Heisenberg visited Holland in 1943, he said that history legitimized Germany's rule over Europe and the world. Casimir reported that Heisenberg had been aware of the German concentration camps and the looting of other countries, but he nevertheless wanted his country to control Europe.

Heisenberg justified his position to Casimir by arguing that only a nation that ruled ruthlessly could maintain itself. Democracy was too weak to rule Europe. Therefore, in Heisenberg's opinion, it was a contest between Germany and Russia. Heisenberg, a pronounced anti-Communist, betrayed his great insensitivity to the plight of his colleagues in occupied Europe by making harsh statements. He coldly drew the logical conclusion from his own arguments, that "a Europe under German leadership might well be the lesser evil."[623] Heisenberg's Dutch colleagues did not appreciate the obtuse message that he gave them, that Germany had to win the war; nor could he understand how or why he had alienated them. He believed that his visit to Holland had gone well, despite all the politics.[624]

Heisenberg had been asked by his Dutch colleagues to visit their country in order to improve their working conditions. This is exactly what he did. On Heisenberg's intervention, Rosenfeld received permission to visit his mother in Belgium.[625] After Heisenberg's visit, the German occupation authorities suddenly announced that the Dutch scientists might be allowed to retain some scientific instruments vital to their research. Kramers and his colleagues immediately submitted a modest list of apparatus they wished to keep. A German official visited Kramers, mentioned that he had spoken with Heisenberg in Berlin, and expressed surprise that the Leyden Laboratory was still closed. This official ostentatiously lifted the ban on research and promised that the Dutch physicists would be told as soon as possible what equipment would not be removed. Heisenberg's Dutch colleagues were sincerely grateful to him.[626]

The German occupation authorities had asked Heisenberg how his visit might be extended and the cultural cooperation between Dutch and German scientists increased. For a long time, he felt unable to answer, but at last gave an apolitical response. Given the state of the war, which was steadily deteriorating for Germany, further visits did not appear to him to be a good idea. He counseled the occupation authorities to wait patiently. But Heisenberg also noted that he considered his trip to have been a success, since it had reopened channels of scientific communication between Dutch physicists and him. His recent correspondence with Kramers had been very valuable. Heisenberg told his countrymen in Holland that he was convinced that scientific relations between the Germans and the Dutch would resume very quickly once the war had come to a happy end.[627]

A little more than a month after returning to Germany from Holland, Heisenberg went east to speak at the German Institute for Eastern Work.[628] Coblitz submitted a second petition in the spring of 1943, and this time it was approved. The ministry made so prompt a decision, and informed Heisenberg so quickly,[629] that he could tell Coblitz of his willingness to speak in the General Government[630] even before the director of the German Institute for Eastern Work had sent him an official invitation.[631]

Around the same time, Heisenberg received recognition from the east of his enhanced professional prestige in another form, the "Copernicus Prize" for excellence in physics. This prize, originally awarded by the University of Königsberg, was now awarded jointly by the university and Frank's institute.[632] Both Heisenberg and Gustav Borger, a Party official from the University Teachers League, saw this honor as yet another blow against the forces of Lenard and Stark. Borger sent Heisenberg his hearty congratulations, since this award represented yet another gratifying official recognition of Heisenberg's work and thereby of theoretical physics.[633] Heisenberg replied that this prize especially pleased him, because it could be interpreted as an official rehabilitation of theoretical physics.[634] As Germany's position in the war grew worse, Heisenberg's prestige as a scientist in Germany rose higher and higher.

Coblitz took care to remind the "in-house physicist" at the German Institute for Eastern Work to attend Heisenberg's lecture, especially since Frank, who was a "close friend" of Heisenberg, had personally invited him.[635] Heisenberg's visit to Krakow was delayed until the end of the year. Frank was either busy or on vacation.[636] Heisenberg had to wait until the dates of his trip to Holland were set in October.[637] A month later, he fell ill.[638] He finally delivered his lecture in the second week of December, only a few months after the German authorities had begun to annihilate the Jewish ghettos in Krakow, Warsaw, and Lodz.[639]

There is no record of how or whether Heisenberg reacted to the razing of the ghettos, but he probably knew that it was happening. Similarly, Heisenberg knew that throughout Europe Germans were pillaging occupied countries and deporting their Jews to concentration camps. But Heisenberg was hardly alone. Every German with eyes to see and ears to hear knew about the concentration camps and that the Jews had vanished from Germany. After the war, many people inside and outside of Germany assumed that Germans like Heisenberg knew about the Holocaust, but nevertheless either did nothing, or even worse, continued to work for the National Socialists. Is this criticism fair?

Philippe Burrin's analysis[640] of the decision to launch the Holocaust helps put Heisenberg's activities into context. According to Burrin, Hitler was torn by two conflicting, if both malevolent, intentions towards the Jews. On one hand, Hitler wanted to purge them from Germany. This goal did not necessarily require genocide, for Hitler and the National Socialist leadership spent a great deal of time and effort on plans to deport Jews to a "reservation" like Madagascar or a region deep in Asiatic Russia. On the other hand, Hitler also wanted to use some Jews as hostages against the international Jewish conspiracy he saw threatening him, his movement, and the German people.

Obviously Hitler could not both eliminate the Jews from the German sphere of influence and simultaneously hold them as hostages. Thus his policy toward Jews vacillated during the first nine years of the Third Reich. His decision to forego both options in order to murder the Jews was the result of a third theme in his irrational worldview. The National Socialist leader blamed the Jews, both inside and outside of Germany, for the German defeat in World War I. As Burrin demonstrates, Hitler consistently threatened the Jews with physical extermination if there was a repeat of World War I, in other words, if "the Jews" once again threatened to betray and defeat Germany.

In the late summer of 1941, it became clear to the German military leadership that the conflict with the Soviet Union would be a long difficult affair, and that ultimately the United States would enter the war on the side of Great Britain. World War II was thereby transformed from the quick painless lightning war to a world-wide war of attrition similar to the conflict Germany lost in 1918. Hitler now ordered a sudden and definitive change in his policy towards the Jews. Emigration, which had been encouraged, was now stopped. Plans for a Jewish reservation were dropped. The uncoordinated murder of Jews by special SS forces in the occupied regions of the Soviet Union was transformed into a systematic, efficient, bureaucratic genocide.

Five or six million Jews were murdered, many killed in gas chambers after being shipped to death camps in overcrowded cattle cars. The Jews were not the only victims of the National Socialists. Another nine or ten million people were starved, shot, or overworked. The National Socialists treated Gypsies like the Jews and murdered forty percent of the one million Gypsies in Europe. Around four million Slavs lost their lives as slave laborers in Germany. Finally, the Germans deliberately allowed two or three million Soviet prisoners of war to die in captivity.[641]

It hardly seems fair to accuse Heisenberg or anyone else of responsibility for the Holocaust before the National Socialist leadership itself decided to commit genocide. Thus Heisenberg's appeal to the SS for a political rehabilitation, his willingness to travel abroad as a goodwill ambassador for National Socialist Germany, and his participation in the wartime German "uranium project"[642] —in other words, his decision to remain in Germany and work within the system—all happened or began before the Holocaust became inevitable. However, Heisenberg knew he was working for a ruthless, racist, and murderous state.

Moreover, Heisenberg did not stop working on nuclear fission, traveling abroad, or enjoying the political backing of patrons in the Third Reich once he learned of the rape of Europe, the deportation of Jews, the razing of the ghettos, or of the death camps. That would have meant taking a clear, courageous, and potentially dangerous stand against National Socialism, something Heisenberg did not do. However, it hardly seems fair to blame Heisenberg for the Holocaust. His conduct was consistent over the course of the Third Reich. It was Hitler who changed his mind.

✠   ✠   ✠

Copenhagen in 1944 During the winter of 1943–1944 the war entered its last, and for the majority of Germans, most hopeless phase. The steady deterioration of German society, including the destruction of cities from the air, interruptions in the transportation system, and increasing shortages of basic necessities, hampered, but did not stop Heisenberg's guest lectures. He did not go to the GCI in Bucharest[643] or to the "German Academy" in Klagenfurt.[644] Instead he stayed in Berlin for the 1944 summer semester to lecture at the university.[645] But he did go to Copenhagen.[646] Heisenberg learned in January that the German occupation authorities had occupied the Bohr Institute. Jørgen Bøggild, the Danish physicist who had been left in charge after Bohr and the Jewish or partly Jewish members had been forced to flee Denmark for Sweden, had been arrested and accused of working with Germany's enemies.

Once the remaining physicists at the Bohr institute realized that their German colleagues had not been responsible for the German takeover, they decided to alert Heisenberg to the occupation and asked the physical chemist Hans Suess—who was passing through Copenhagen on his way south from Norway—to pass on the message. Heisenberg learned of the occupation from Suess on 5 January 1944 and arranged to be part of the German commission that would investigate whether the research at the Bohr Institute had been contributing to the Allied war effort.[647]

Von Weizsäcker found out to his dismay that the German officials in Copenhagen were considering making him the new director of Bohr's old institute. He did not want to confront his Danish colleagues as a conqueror and asked Heisenberg to use his influence to kill the plan.[648] In the company of the Army physicist

Kurt Diebner and others, Heisenberg traveled to Denmark on 24 January and met with the plenipotentiary of the German Reich in Copenhagen.[649] The German authorities were debating whether to staff the Bohr institute with German physicists, to force the Danish scientists at this institute to contribute to the German war effort, or to strip the institute of all equipment needed in Germany.[650]

Heisenberg obviously wanted to arrange as beneficial a settlement as possible for the Danes. He toured the high-voltage equipment and the cyclotron at the institution with some occupation officials, emphasizing how complicated the equipment was and how difficult to move. The next day, the German authorities informed the Danish Foreign Office that the Bohr institute would be reopened without conditions and released Bøggild.[651] Heisenberg subsequently told Johannes Jensen, a colleague who had many friends and acquaintances at the Bohr institute, that the Danes were very happy about this outcome.[652]

A month after his visit to Denmark, Heisenberg received an invitation by way of the Foreign Office and the German occupation officials to speak again at the Copenhagen GCI.[653] Heisenberg accordingly spent four days in Denmark, April 18 to 22, as guest of Otto Höfler, the new director of the GCI. On the evening of 19 April, Heisenberg gave his talk, "The smallest building blocks of matter," before an audience made up almost completely of Germans. Heisenberg's Danish colleagues refused to attend, including the scientists who had attended the 1941 astrophysics conference and who, until the resignation of the Danish government, had participated in the programs of the GCI. The following day Heisenberg had lunch with the plenipotentiary of the Reich, Dr. Best, and spent the evening as Höfler's guest with several representatives of cultural politics in Scandinavia.

On 19 April, Heisenberg also paid a visit to Bohr's old institute, whereupon Heisenberg's Danish colleagues invited him to give them a talk on his own work. Heisenberg subsequently met with several Danish colleagues and their wives as the guest of Professor Møller. On 21 April Heisenberg lectured on "the theory of elementary particles" in Danish, followed by a brief institute tea.

Heisenberg asked the Danes why they had not come to his talk at the GCI. They replied that, because of the tense political relationship that had existed between Germany and Denmark since the Danish government resigned in 1943, they wanted nothing to do with the political GCI. After discussing all this information in his report, Heisenberg went on to support energetically what the director of the Copenhagen GCI had told him: Höfler would never be able to win over the Danes and gain their cooperation unless, for the time being, he restricted himself to purely scientific and scholarly work. The side of his work that had more to do with propaganda, such as guest lectures and the like, should be postponed to a later, more opportune time. Heisenberg closed his report with the same conviction he had expressed after his last trip to Holland: once the war had come to a happy end, scientific cooperation with the Danes would not be difficult.[654] Indeed, after the war Heisenberg had a great deal of difficulty understanding why he had alienated his foreign colleagues.

The director of the Copenhagen institute during the last years of the Third Reich may have been typical of the scholars sent as cultural emissaries to foreign countries by the National Socialist state. Höfler's specialty was Germanic philology. He had spent many years in Scandinavia and had taught at the University of Uppsala. The Copenhagen GCI did not limit its activities to Denmark, but attempted to influence cultural policy in Sweden and Norway as well. He had connections with Scandinavian colleagues, knew the countries, and spoke the languages.[655]

Shortly after Höfler joined the NSDAP in the spring of 1937,[656] he was appointed to the Research Council of the "Ahnenerbe," a branch of the SS.[657] The Ahnenerbe supported a wide range of research. Some topics would now be considered unscientific or even pseudo-science, such the "World Ice Theory" developed in the early twentieth century by Hanns Hörbiger. Both Himmler and Hitler were very interested in Hörbiger's work, which argued that the universe was composed of ice.[658] The Ahnenerbe also sponsored respectable science, such as entomology and plant genetics. Finally, the Ahnenerbe was the branch of the

SS which planned, financed, and carried out inhuman experiments with prisoners of war and concentration camp inmates.[659]

In 1938, the SS helped Höfler trade his professorship at Kiel for a more prestigious one at Munich, and in return he placed his expertise in Germanic philology and close connections in Scandinavia at the service of the Ahnenerbe's efforts to use the field of Germanic prehistory in order to justify the dominance of the "Aryan" race.[660] In 1942, before moving from the University of Munich to the Copenhagen GCI, Höfler visited Denmark with SS papers to do research the SS characterized as intelligence work.[661]

After the second world war Höfler applied for a teaching position at the University of Munich. A university official asked Heisenberg in 1949 whether Höfler had strictly limited himself to scholarship while in Copenhagen, or had engaged in cultural propaganda.[662] Heisenberg's evasive answer[663] provides insight into his perception and continued support of the GCIs. First, he claimed that he had never met Höfler personally. Perhaps he had forgotten about his 1944 meeting with Höfler,[664] of which Höfler reminded him in 1947.[665]

Next Heisenberg asserted that the Copenhagen GCI had not had an entirely bad reputation and that it had not been a source of explicit National Socialist propaganda. If the Danes had stopped frequenting the GCI, that was not Höfler's fault. They had merely concluded that the Germans would lose the war. Heisenberg said that he had never heard criticism of Höfler by the Danes, although he did admit that the Danish scientists would hardly have expressed such complaints to him. Heisenberg closed his report on Höfler by noting that even if the latter had not been as successful as Freyer, the president of the Budapest GCI, Höfler had not left a negative impression behind in Denmark.[666]

As late as 1949, Heisenberg had few misgivings either about his past associations with, or the goals of, these institutes. Heisenberg may have been unaware of Höfler's connections to the SS, but that would hardly explain the physicist's participation in National Socialist cultural propaganda. From Heisenberg's perspective, the GCIs afforded him the opportunity of retaining contact with colleagues all over Europe. A boycott of them would have done him no good, nor would it have benefited German physicists or scholars in other countries.

After the war Heisenberg wrestled with this dilemma in a memo entitled "The active and passive opposition in the Third Reich." This essay—apparently never published or circulated—offers a unique opportunity to get inside Heisenberg's mind and arguably demonstrates his postwar denial of the true nature of both the Third Reich and the role he played in it.

If the vast majority of Germans had refused any collaboration with National Socialism in 1933, Heisenberg noted, then much misfortune would have been avoided. But that did not happen. Rather, the National Socialist system had understood how to win the support of the masses. Once the National Socialists had gained control of the government, the relatively thin layer of people whose certain instinct told them that the new system was bad from the ground up, now only had the opportunity of "passive" or "active" opposition.

Heisenberg noted that, on one hand, these people could have condemned the National Socialist system as basically bad and a threat to Germany and Europe, but concluded, nonetheless, that there was nothing that could be done. Whoever reasoned that way could either emigrate or deny responsibility, and wait until the system was overcome from the outside. Heisenberg designated this behavior as "passive" opposition. Another group, he went on, judged the situation as follows. A war, even if it served to overthrow National Socialism, was such a horrible catastrophe, and would cost so many millions of people their lives, that everything had to be done to avoid it or to reduce its horror. Many people who thought so, but did not comprehend the stability of a modern dictatorship, tried the path of open, immediate resistance during the first years and ended up in concentration camps.

For others, Heisenberg added, individuals who recognized the hopelessness of a direct attack on the dictatorship, the only path remaining was the acquisition or preservation of a certain amount of influence. Such people risked being branded collaborators. Heisenberg now argued that this course was the only way to bring about change in National Socialism and described it as "active" opposition. This position was much more difficult and ambiguous than passive opposition, since the activist had to make concessions at unimportant places in order to be able to influence important matters.[667]

Heisenberg's retrospective portrayal of "active" and "passive" opposition during the Third Reich makes clear what he chose to believe after the war. By staying and working within the National Socialist system, accepting responsibility and thereby being in a position to wield influence, Heisenberg had "actively" opposed Hitler.

Heisenberg's last foreign lectures took place in Geneva and Zurich in the autumn of 1944.[668] When he met with his Swiss colleagues, Heisenberg repeated what he had told their Dutch counterparts a year before: only Germany stood between Russia and European civilization.[669] Furthermore, when Heisenberg was asked about the prospects for a German victory in Europe, he said that it would have been nice if Germany had won.[670]

This answer did not please either the Swiss or the Germans. The former would assume that Heisenberg wanted National Socialism to dominate Europe, if not the world. The latter would consider Heisenberg's comment defeatism, something which became a serious offense during the last, terrible months of the war. Finally, Heisenberg's remark need not have been a conscious endorsement of National Socialism. Once the war began, many Germans separated in their own minds their support of Germany from that of Hitler's movement. This self-deluding distinction was important, for it allowed the National Socialist state to harness the energies of the many Germans who did not support Hitler, but also wanted Germany to win the war.

Significantly, Heisenberg never got around to sending in a report on his 1944 trip to Switzerland. In late March 1945 REM reminded him of his omission,[671] but by this time Heisenberg was more concerned about the advancing American forces than about the bureaucrats in Berlin.

Foreign scientists have shown a great deal of ambivalence toward Heisenberg and von Weizsäcker since the end of World War II.[672] This ambivalence derives largely from the talks the two German physicists gave in foreign countries during the war as well as the postwar apologia they have used to justify their conduct in the Third Reich. But Heisenberg and von Weizsäcker did not merely participate in National Socialist cultural propaganda. They were also exploited by Hitler and his followers, as were many Germans.

Heisenberg never spread vulgar National Socialist propaganda. Even his comments to Casimir were couched in terms of Germany, not Hitler's movement. Every one of Heisenberg's official visits was restricted to scientific talks. But that was precisely what the National Socialist officials responsible for cultural propaganda wanted him to do as part of an effective division of labor. Heisenberg represented the "better side" of National Socialist Germany as a "good German," an apolitical Nobel laureate willing to serve as a goodwill ambassador for German culture while other Germans were invading, occupying, exploiting, and sometimes murderously ravaging the very same countries.

The German Cultural Institutes and comparable institutions such as the German Institute for Eastern Work provide vivid examples for the distortion and abuse of science and culture. In the eyes of many native scientists, these institutes were centers of scientific and cultural collaboration with National Socialism as well as symbols of the German occupation and exploitation of their homeland. As long as he could lecture in German, Heisenberg accepted all offers of speaking engagements at such institutes and thereby alienated and deeply disappointed many of his foreign colleagues.

Heisenberg was either unable to understand or unwilling to confront the cause and effect of this alienation. By delivering lectures there, he supported and thereby legitimated the National Socialist policy of cultural propaganda. When he could, he aided foreign colleagues in trouble, including Jewish scientists. He did this at considerable risk to himself, and his colleagues were grateful. But this gratitude could not make up for the alienation caused by his participation in cultural propaganda and his personal identification with the German war effort and German armed forces.

The National Socialist state reexamined its policy toward modern physics during the course of the Third Reich and especially during the war, with the result that the irrational and barren *Deutsche Physik* was eventually discarded in favor of modern physics, with its recognized economic and military utility. But it was first and foremost Heisenberg, and not modern physics, that came under dangerous political and ideological attack in the *Völkischer Beobachter* and *Das Schwarze Korps*, and it was first and foremost Heisenberg, not the theory of relativity or quantum physics, who emerged victorious with a political rehabilitation and enhanced prestige.

The SS report on Heisenberg suggests that scientific arguments alone did not win this battle. Industrial scientists and researchers with close ties to the armed forces played a crucial role. The SS and the Party accepted the judgment of Ludwig Prandtl and Carl Ramsauer, that modern physics was useful and needed support, and found a politically and ideologically acceptable justification for its rehabilitation. Heisenberg's appeal as a "good German" and especially his long-standing association with the armed forces made it easier for the National Socialist state to accept his physics. Once the ideological taint had been removed from modern physics, Heisenberg could be also used as a cultural propaganda tool.

The political rehabilitation of Heisenberg was necessary before the National Socialist state could take full advantage of his propaganda value. For Heisenberg to be useful in a cultural propaganda sense—or for that matter, to be useful in the training of physicists or for research—he had to be used; for him to be used, he had, to some degree, to be trusted; for him to be trusted, the National Socialist state had to make some concessions with respect to the ideological purity of physics. The very utilitarian and international character of modern physics was used to facilitate cultural cooperation and ultimately collaboration between scientists in foreign countries and National Socialism.

Finally, Heisenberg's foreign lectures illuminate the problematic black-and-white dichotomy of resistance versus collaboration. Heisenberg's 1941 visit to Copenhagen has been portrayed as proof that either: (1) the physicist willingly collaborated with the "Nazis" to exploit Bohr; or (2) he resisted Hitler by warning the Allies of the German atom bomb.[673] When this visit is seen in context, it is clear that the truth lies somewhere in the middle.

Carl Friedrich von Weizsäcker has insisted that *intent*, not *action*, is most important. He and Heisenberg traveled to Copenhagen in order to help their Danish colleague Niels Bohr. But what kind of help did Heisenberg and von Weizsäcker offer Bohr in the fall of 1941, when German victory appeared certain? They urged him to cooperate with the German authorities and especially the German Cultural Institute in Copenhagen. Today it is clear that this was bad advice; at that time it may not have been so clear. It is hardly surprising that Heisenberg and von Weizsäcker offered Bohr precisely this advice. They merely advised Bohr to do what they were doing.

8

Hitler's Bomb

The 1941 meeting between Bohr and Heisenberg is controversial because it is part of the debates surrounding "Hitler's Bomb." During the war both Germany and the United States investigated the economic and military potential of applied nuclear fission. The American effort, otherwise known as the Manhattan Project, built the bombs which fell on Hiroshima and Nagasaki. Obviously the Germans did not manufacture nuclear weapons before Germany surrendered. But ever since the end of the war, scientists and non-scientists both inside and outside of Germany have argued over *why* the Germans failed, and whether the word failure is an appropriate description. This chapter will survey the German uranium project in the context of science under National Socialism.

✠   ✠   ✠

Physics and Politics in Weimar Germany (1919–1932) Today it is clear that science in general and physics in particular can be politicized, but science has not always been so susceptible to external influences. An irreversible politicization of science took

183

place in Germany during the first half of the twentieth century, beginning with the exceptional publicity given to Albert Einstein's theory of relativity and ending with the race for nuclear weapons. Although physics had been temporarily politicized at different times and in different places, since 1945 governments have seen this science as a potential source of political power.

Einstein was a respected scientist even before World War I. But the unusual popularity his theory of relativity enjoyed during and after the war, combined with his unconventional personal style and political stance, transformed him into a cultural and political icon during the Weimar Republic. The experimental verification of relativity in 1919 and the subsequent public fascination, if not obsession with Einstein made the pacifist, democrat, and Jew a cultural and political symbol that transcended his physics and incurred the wrath of both political conservatives and scientific opponents. This political and scientific opposition to Einstein and his theory of relativity created an ideological struggle between "Aryan" and "Jewish" physics during the Third Reich.[674]

Opposition to Einstein and modern physics was fueled by the political and economic aftermath of World War I in Germany. The lost war was a catastrophe for the conservative majority of academic scientists. They often reacted by asserting that science and scholarship were all that Germany had left as a world power,[675] an attitude which accelerated and deepened the politicization of physics.

The weak economy and hyper-inflation ruined the endowments of many scientific institutions—not to mention the savings of scientists—and forced researchers to compete for the ever-shrinking amount of available funding and to become more dependent on the generosity of the central government and German industry. This shortage of funds forced the scientific community to work with the government to create the modern peer-review system of science funding. Institutions like the governmental Emergency Foundation for German Science and the private Helmholtz Foundation relied on expert committees to decide which scientists would receive support.[676]

A small group of senior German physicists like Max Planck dominated the expert committees within the peer review system and thereby influenced, if not controlled, which research was funded. The major beneficiaries of this system included the creators of quantum mechanics, including Max Born, Werner Heisenberg, Pascual Jordan, and Erwin Schrödinger. In contrast, the conservative scientists who rejected modern physics did not have a large share in the new funding system.

Perhaps most important, the politically conservative scientists who opposed the Weimar Republic and rejected the political stance of liberal colleagues like Einstein were often the same researchers who were unable or unwilling to accept quantum mechanics and relativity.[677] Similarly, Einstein's non-scientific political opponents used his controversial theory of relativity as a means to attack him. Einstein's physics and politics thus merged into a single target for political and scientific conservatives. The political and economic upheaval following Germany's defeat thus made modern physics—roughly speaking quantum mechanics and relativity theory—at once both the pride of German science and the target of scientists and laymen who opposed a liberal, democratic worldview.

Two German physicists and Nobel laureates, Philipp Lenard and Johannes Stark, vigorously opposed the Weimar Republic, and felt betrayed by the lack of recognition given to them by their colleagues and government. They were professionally opposed to (in each case, different) elements of modern physics. Such sentiments were common in Germany between the wars, but they went further. By 1933 both scientists were channeling their personal and professional discontent into the virulent anti-Semitism so common on the political right and public support of Adolf Hitler. When the National Socialists came to power in 1933, Lenard and Stark gained access to political power and influential friends in the new regime.

⊞   ⊞   ⊞

Nazification and Militarization (1933–1939) When the Allies defeated the Third Reich and the National Socialist leadership was dead or being tried for war crimes, there was a general consensus outside of Germany that the German people had to be "denazified." But if the Germans had to be denazified after 1945, then they must also have been nazified sometime between 1933 and the end of the war. Nazification can be defined as follows: the effective, significant, and conscious collaboration with most—but not necessarily all—of National Socialist policy. Since the attitudes, assumptions, and actions of German scientists varied greatly during the Third Reich, so did the form and course of their interactions with National Socialism.

For German politics, 1932 was a tumultuous year. Adolf Hitler's National Socialist German Workers Party had emerged from obscurity to become the largest political party in Germany. Ironically, when German President Paul von Hindenburg appointed Hitler Reich Chancellor in January 1933, the National Socialists were on the way down; they had peaked the previous year and were struggling to hold their political movement together.

Hitler had been helped into power by an intriguing circle of industrialists, aristocrats, and senior military officers who hoped to use the National Socialist leader for their own ends. Hitler proved to be the more skillful politician and exploited the collaboration of Germany's old elites to help his radical, racist, and ruthless movement eliminate step-by-step all opposition during the first few years of the Third Reich. The old elites retained a little autonomy until the eve of World War II, when Hitler purged the Army leadership. Personal scandals were exploited or manufactured for Field Marshall von Blomberg and Army Commander-in-Chief General von Fritsch, two officers who had expressed concern that Germany was not yet ready to fight. They were eased out of their posts and replaced by more pliable men. In addition, fourteen senior generals were retired and forty-six others required to change their commands. Hitler personally took over as Commander-in-Chief of the Armed Forces.[678]

Both the purge of the German civil service[679] and of German science at the start of the Third Reich are well known.[680] The so-called seizure of power[681] by the National Socialists dramatically and decisively affected all parts of German society, including science. But both scientists and historians of science have sometimes failed to recognize that the purge of scientists was not a conscious National Socialist policy against science *in particular* and, at least for academics, was an automatic result of the greater civil service purge.

The National Socialist leadership was hardly concerned enough about any particular science, or even science itself, to single it out for special treatment. Education in general and university education in particular were priorities for Germany's new rulers, but in this regard physicists were treated no differently from their non-scientific colleagues.

Albert Einstein, perhaps the most famous scientist purged by the National Socialists, represents the exception that proves the rule. Hitler's movement singled out Einstein for wrathful special treatment precisely because his public stature represented a real political threat. However, the thorough and ruthless purge of the civil service effectively "cleansed" the universities and state-funded research institutions (like the Kaiser Wilhelm Society) of Jewish, leftist, and other elements incompatible with the new Germany, thereby striking a heavy blow to all branches of German science.

It is important to recognize *how* the National Socialist purge and reorganization of German society functioned, *why* it was successful, and what *pattern* it followed. First, Hitler and his followers needed and received assistance from influential members of Germany's conservative elites—including scientists. Second, and most important for this subject, the purge was neither centrally planned, coordinated, nor implemented. Instead the seizure of power was characterized by uncoordinated and often unsolicited pressure from National Socialist rank-and-file party members and SA. This violent and often unsolicited pressure was then exploited by the National Socialist authorities to eliminate all opposition.[682]

Such unsolicited, yet often welcome, attacks from below by the masses making up the basis of Hitler's movement were often subsequently used by the National Socialist government to justify further repression from above by blaming the victims for inciting the violence.[683] But since the National Socialist leadership also wished to present an image of a peaceful, orderly society under their control, such "revolution from below" eventually became counterproductive. On 6 July 1933 Hitler publicly called for "evolution, not revolution," a thinly veiled threat to his own followers.[684] When the SA leadership persisted in its calls for a "second revolution" which would have benefited in particular the lower levels of the National Socialist movement, Hitler purged his own movement.

In the summer of 1934 German President Otto von Hindenburg, one of the few remaining checks on Hitler's government, was dying. Hitler intended to merge the office of president into his own position of Chancellor, but that required the blessing of the Armed Forces, the only remaining part of the German state which could launch a putsch against him. The Army feared the SA, and with good reason. Its leadership wanted to turn the SA into a political army and to absorb the armed forces in the process. The SS was also involved, because it technically was still a subsidiary of the SA and wanted greater independence. Pressure on Hitler from the Army leadership and the SS finally forced his hand. On 30 June Hitler personally supervised the arrest of Ernst Röhm and the majority of the SA leadership. Most were subsequently murdered by the SS with Army logistical support. This bloody "night of the long knives" permanently silenced calls for a second revolution.[685]

The nazification of German science in general and physics in particular followed this SA model and its four stages, although recalcitrant scientists were disciplined, not murdered: (1) revolution from below, uncoordinated and unsolicited attacks in the name of National Socialism; (2) evolution, not revolution, the National Socialist government orders that henceforth all change will be directed by the responsible authorities or occur through official channels; (3) second revolution, the National Socialist rank-and-file nevertheless continues its agitation; and (4) finally the National Socialist revolution devouring its own children, purging or disciplining its undisciplined followers.[686] The physics equivalent of the SA was the *Deutsche Physik* movement, which called for a more "Aryan" and less "Jewish" science.[687]

The followers of Lenard and Stark wanted to achieve a second revolution in German physics which would go beyond the initial purge of the civil service and would ensure that they would henceforth receive the best university appointments. Their weapon was a very effective campaign of character assassination. However, by 1936 Stark and his allies were beginning to get in the way of other, more influential forces within the National Socialist movement, including officials within the Ministry of Education and the leadership of the SS.

Deutsche Physik was first opposed and then neutralized by other and stronger parts of the National Socialist movement because the long-standing goals of the former conflicted with the new ambitions of the latter. In contrast to the scientifically sterile *Deutsche Physik*, the established physics community could and did effectively contribute to rearmament and the war effort by training scientists, engineers, and technicians for the armament industry as well as developing new weapons and industrial processes.

This increase in German military strength and initial military successes in turn increased public support for the Third Reich and facilitated the most extreme and murderous National Socialist policies: the creation of a racially pure society in Germany; cultural imperialism; geographic expansion through military aggression; and finally genocide. Although the *Deutsche Physik* movement failed in its efforts to make German physics more National Socialist by attacking modern physics and certain physicists, ironically the successful struggle by the established physics community against *Deutsche Physik* and the consequential collaboration with the National Socialist state it entailed did.

Why was German physics nazified in this way? The adherents of *Deutsche Physik* simply tried to expand their influence within the German physics community any way they could, and initially their strategy appeared successful. The established German physics community could easily find influential and sympathetic patrons within the sometimes chaotic and contradictory political structure of the Third Reich. This support was sometimes given for reasons of principle, sometimes as a cynical, tactical stance within the shifting politics of the National Socialist state, but no matter why these patrons chose to side against *Deutsche Physik*, some of them were in a very strong position to do so.

But why did the overwhelming majority of German physicists ally themselves with, or submit themselves to forces within the Third Reich and portions of National Socialist policy? Obviously because when compared to the ideological threat represented by *Deutsche Physik*, this course seemed less objectionable because it would provide more professional autonomy. However, this apparent gain in autonomy was misleading. The established physics community had rid itself of *Deutsche Physik*, but now had to demonstrate both loyalty and usefulness to the Third Reich. One of the most controversial and potentially dangerous collaborations between German physicists and the National Socialist state was the uranium project, research into the military and economic applications of nuclear fission.

✠   ✠   ✠

Nuclear Fission (November 1938–August 1939) Otto Hahn and Fritz Strassmann, two chemists working at the Berlin Kaiser Wilhelm Institute for Chemistry, made a discovery in late 1938 which, in time, changed the world. When they bombarded uranium, the heaviest natural element, with neutrons (nuclear particles without charge but with mass) they found barium, an element half the mass of uranium. Their Jewish physicist colleague Lise Meitner helped make the initial discovery possible, but had fled Germany earlier in 1938 after the Third Reich had absorbed Austria, which ended the protection her Austrian passport had once provided.

When news of Hahn's and Strassmann's striking result reached Meitner in Sweden, she encouraged her former colleagues. When she subsequently met her physicist nephew Otto Frisch in Denmark, they solved the riddle together: the uranium nucleus had split in two like a liquid drop. Although Frisch and Meitner were among the first scientists to extend the Berlin results, it is perhaps more significant and important that so many different researchers in different countries carried out the same experiments, achieved the same results, and came to the same conclusions: when uranium nuclei split, they released both energy and more neutrons.

Scientists around the world took up this research immediately and raced to be the first to explain, expand, and apply this phenomenon. Personal and professional ambition as well as the obvious potential of nuclear fission ensured that long before scientists began withholding their results in the shadow of World War II, their publications had already demonstrated that uranium fission released great amounts of energy as well as enough neutrons to make possible energy-producing and exponentially increasing nuclear fission chain reactions.

It was only a very short step from these results to the realization that nuclear fission had consequential economic and military applications: a controlled chain reaction could be used to generate electricity; an uncontrolled chain reaction would represent a powerful new explosive. Scientists went to the responsible military authorities in almost every country and passed on the same message, that it might be possible to harness nuclear fission both as *nuclear* explosives of hitherto unknown power and as nuclear energy. They noted that enemy countries were probably already working on uranium; the government had to support a research program in order to determine whether nuclear weapons could be built, how they would be built, and whether they should be built.

Even though researchers throughout Europe and North America went to their governments with this same message, historians and scientists who have studied "Hitler's Bomb" have often distinguished between the German scientists who enlightened their military and their colleagues in other countries who did the same. While American, French, and British scientists are praised for these efforts, their German counterparts are criticized. Indeed, this distinction plays an important role in the persistent fascination with Hitler's bomb. There is an important difference here, but it is not in what the scientists did, rather in what sort of regime they were serving.

✠   ✠   ✠

Lightning War (September 1939–November 1941) The German uranium project did not progress until after the invasion of Poland in September 1939. These two events were connected. The overwhelming majority of Germans and German scientists rallied to the flag once the war had begun, including many individuals who opposed or at least did not wholeheartedly support National Socialism. War also made it both more attractive and easier for Army Ordnance to become involved with scientific research projects which promised powerful new weapons,

Finally, a fundamental transformation in National Socialists' attitudes towards science and science-based technology began during the middle thirties with rearmament and accelerated with the outbreak of war. Military and economic power took precedence over ideological purity. Scientists who could offer something useful for the war effort could now eclipse their colleagues who were ideologically correct but scientifically inferior.

The quality of the German uranium effort can best be judged when compared to its Allied counterpart. During the Lightning War phase the two projects ran astonishingly parallel. With a few exceptions, the Germans and the Americans examined the same subjects, used the same methods, asked the same questions, and found the same answers. There were many different reasons why German scientists chose to participate in the nuclear power project: scientific interest, careerism, financial and material support, exemptions from military service, patriotism, nationalism, and National Socialism—in other words, with the exception of the last point, the same motivations as in other countries.

But motivation and scientific ability alone do not tell the whole story. The political and military leadership was in *control* of the research and had the power of decision. In Germany it was Army Ordnance, and not the academic scientists. The situation in the United States during the war and in the Soviet Union after the war was no different. The scientists actually carrying out the research could not and did not decide whether the research was begun, whether and how it was continued, and if successful, what would be done with the new weapons once they had been created.

These decisions were made by governmental and military officials. Moreover, in Germany Army Ordnance not only had the power of decision-making, it also had its own competent and loyal scientists, who could well judge the technical and scientific side of the project. The influence of the research scientists over the project, let alone their ability to control it, was limited—although these scientists very often deluded themselves and believed that they were really in charge.

The German uranium research must also be seen in the context of the ever-changing state of the war and, in particular, of the Lightning War which ran from 1939 to the winter of 1941–1942. Germany used the tactic of massive sudden attacks to overwhelm an opponent, strip the conquered country of resources, and use these resources to launch the next attack. The secret reports gathered by the SS Security Service describe how the combination of military success and skillful propaganda combined perpetually to convince most Germans that the war was almost over. Thus in the summer of 1940 it appeared that the war would be over by Christmas. By Christmas it seemed likely that the war would be over by the spring, etc.

Throughout the Lightning War the overwhelming majority of Germans (and most likely German scientists as well) believed that the war would soon end with victory. "Wonder weapons" were not needed. Army Ordnance was in no hurry to have weapons which would not be ready until after the war, and the scientists were under no great pressure to deliver them. Indeed, some of the scientists may well have believed that they were exploiting the Army and National Socialist government for their own ends by receiving both exemptions from military service and research support for something irrelevant to the war being waged.

Postwar claims by project scientists such as Werner Heisenberg that he had been convinced from the very beginning that Hitler would lose the war do not ring true.[688] Heisenberg may well have believed that in September 1939, and it is very likely that after the war he chose to remember his feelings and beliefs in this way. But it is very difficult to fathom that between the summer of 1939 and the autumn of 1941, when German armies inexorably attacked, conquered, and occupied most of Europe, Heisenberg could have believed anything other than what the overwhelming majority of his countrymen did: that the war would soon be over, with a German victory.

It is extremely difficult to judge the motivations of these German scientists during the first phase of the war. Any such judgment should really attempt the impossible and try temporarily to forget the Holocaust that began in the fall of 1941 and the unconditional surrender of German armed forces in the spring of 1945. During the Lightning War these scientists could and did work without great pressure, secure in the knowledge that the war would end with victory before any such nuclear weapons would be needed.

✠   ✠   ✠

The War Slows Down (November 1941–November 1942)
The German offensive ground to a halt short of Moscow in the winter of 1941–1942. The subsequent counterattack by the Soviet Red Army pushed back the German forces for the first time in the war and brought the Lightning War to a definitive end. The Japanese attack on Pearl Harbor and Hitler's subsequent decision to declare war against the United States decisively altered the balance of power and drastically changed the political context for uranium research. The Lightning War was now replaced by a war of attrition, where natural resources, industrial capacity, and manpower would determine the victor as the two sides tried to wear down each other. This was a type of warfare which the United States and the Soviet Union were in the best position to win, not Germany. But even though it was now clear that the war would last much

longer, most Germans still believed that they would eventually be victorious.

The responsible science policy officials in Germany and the United States independently reviewed their respective nuclear fission research programs with one fundamental question in mind: could nuclear weapons be manufactured soon enough to influence the outcome of the war from either side? Since Germany obviously needed a more efficient and better organized war economy, Army Ordnance asked its scientists for the first time whether they could expect nuclear weapons soon. American officials asked their scientists similar questions at almost the same time.

Although the hard scientific results were practically the same on both sides, the political, economic, and ideological perspectives were decisively different. In the United States Vannevar Bush, science policy advisor to President Roosevelt and head of the Office for Scientific Research and Development, decided that nuclear weapons might be produced in time, so that the Americans and their Allies *had* to try. In Germany Erich Schumann, head of the research section of Army Ordnance, decided that neither side could produce nuclear weapons in time, so that the Germans *must not* waste valuable resources and time by trying. But as will be discussed below, the meaning of even the word "try" is not as clear-cut as it might appear.

The response by the German uranium scientists can best be appreciated when compared both to its American counterpart and to the selling of the German rocket effort. The German uranium scientists' report was practically identical to that of their American counterparts. There was a great difference between Berlin and Washington, but it lay in perception of the decision-makers, not science. Whereas the American leadership assumed that it would take four to five years to wear down the Third Reich, German political, industrial, and military leaders reckoned with a war of only two or three years more—win or lose. Thus the same scientific results meant that in the United States nuclear weapons could win the war but in Germany could only divert resources away from the immediate war effort.

Schumann's negative decision on nuclear weapons can be better understood when it is compared with his previous decision in 1939 to support the rocket research of Walter Dornberger and Wernher von Braun. When Schumann asked the nuclear scientists whether atom bombs could be manufactured in time to help win the war, they responded that nuclear weapons were certainly possible in principle, but in practice they would require such huge investments in manpower and resources that they were irrelevant to the conflict Germany was fighting.

In other words, the German uranium scientists never pushed nuclear energy or weapons. Moreover, their caution was very prudent. It was dangerous in the Third Reich to promise what could not be delivered. In contrast, when Schumann asked von Braun and his colleagues whether rockets could influence the outcome of the war, they merely replied that if the authorities would give them enough support, they would succeed.[689]

The different decisions reached in Berlin and Washington had corresponding consequences. The Germans pushed the rocket project to murderous extremes, using slave labor drawn from Soviet prisoners of war and concentration camp inmates. They swallowed up huge resources on the scale of what the Americans invested in the Manhattan Project, which was used ruthlessly if ineffectively against civilians in Belgium and England. The rockets caused terror, but were so inaccurate that they were a strategic failure and a waste of resources. Rockets became an effective weapon only after their accuracy was improved and they were coupled with nuclear weapons.[690]

Similarly, although the nuclear research programs in America and Germany had been comparable up to January 1942, this situation quickly changed. Between January and June 1942, the Americans made the huge and obviously necessary investments of manpower, money, and materials and set off on the road to the atom bomb; the Germans did not. By the summer of 1942 the Americans had accomplished what the Germans had almost, but not quite achieved by the end of the war: a nuclear reactor which could sustain an energy-producing nuclear fission chain reaction and the complete isotope separation of a tiny amount of uranium; in other words, the manufacture of a very small amount of nuclear explosives.

But this stark contrast between the German and Allied achievements should not obscure the fact that the German researchers simply carried on with their research at the laboratory level and continued to investigate all possible applications of nuclear fission, including military uses. In particular, despite the claims to the contrary made by Heisenberg and others after the war, the responsible authorities never made a decision or gave a command henceforth to research and develop only the "peaceful" applications of nuclear fission and make it useful for humankind.[691] Instead Schumann made a "non-decision." The research would not be shifted up to the level obviously necessary for the wartime manufacture of nuclear weapons, but the research program would also continue without change or interruption. Everyone agreed that the great future potential of nuclear fission justified further research, even if it would not decide the war.

Heisenberg's postwar claims that he and his colleagues had kept control of the research in their hands were either disingenuous or at best naive.[692] There is one compelling explanation why Heisenberg and some of his colleagues chose to exaggerate and misrepresent the amount of influence they held over the German uranium project: before they could claim that they had resisted Hitler by denying him nuclear weapons, they first had to convince their listeners that they had been in control.

✠   ✠   ✠

The War Is Lost (November 1942–April 1945) The German catastrophe at Stalingrad decisively altered the German position yet again and simultaneously began the period of wonder weapons. German forces had captured Stalingrad with a great deal of effort, but were soon put on the defensive when the Red Army counterattacked and encircled the city. Although the German forces could have broken out, Hitler ordered them to stay and fight. After his men suffered a high rate of casualties, the German commander nevertheless surrendered and took his remaining men into Soviet captivity. Very few of them ever returned to Germany.[693]

The surrender of the German forces shattered the myth of Hitler's infallibility. Perhaps most important, Josef Goebbels' propaganda machine, which had continued to claim that the conflict was going well, was forced to announce the "hero's death" of hundreds of thousands of troops and suffered an irreparable loss of credibility. After Stalingrad the first real doubts about the outcome of the war took root in the German population, and most probably also among German scientists. These doubts were starkly reinforced by the continual deterioration of the war, as the front receded and the Allied bombing of Germany began in earnest.

The worse the war became, the louder and more desperate the search for wonder weapons which could turn the apparent defeat into sudden victory. Ironically, applied nuclear fission was one of the few recent scientific discoveries that were not considered. That possibility had already been investigated and discarded. Despite the ever-worsening state of the war, the bombing attacks that destroyed their institutes and threatened their lives, etc., the uranium scientists continued working with ever greater, if not desperate efforts.

There was no hint of defeatism, rather an enhanced determination to reach their relatively modest goals: building a nuclear reactor which could sustain a controlled chain reaction, and separating out small amounts of uranium isotope 235, a nuclear explosive. Ironically, the German scientists involved with uranium assumed that they were ahead of their rivals in other countries in the race to harness nuclear fission. For them reaching their goal was also being the *first* to do so, an accomplishment which would have obvious professional rewards, no matter who won the war.

Moreover, the very goals of the German uranium project changed over time. Once Army Ordnance had effectively frozen the program at the laboratory level, the progress of the research was limited by the immediate effects of the war: scientists were called up; laboratories were destroyed by bombs; materials and apparatus were in short supply or unavailable; and the scientists were forced to evacuate from the larger German cities to the relatively peaceful countryside. From the fall of Stalingrad to the end of the war the modest goal of the uranium project was to build a nuclear reactor which could sustain a nuclear fission chain reaction for a significant amount of time and to achieve the complete separation of at least tiny amounts of the uranium isotopes.

The threat of impending doom also provoked a perhaps natural human reaction among the scientists to lower their heads and bury themselves in their research. The closer the bombs and fronts came, the harder these scientists worked. By now none of them believed that their work could bring a German victory, although a few administrators did flirt with disaster by dangling such prospects before prospective patrons in the National Socialist state. Thus work on applied nuclear fission in Germany had none of the moral overtones which appeared in the United States after the successful atom bomb test in the New Mexico desert and everywhere else after the attack on Hiroshima. Moreover, the postwar claims by Heisenberg and others, that this moral question dominated their thinking during the war, also do not ring true.[694]

But it is not enough merely to investigate the *scientists'* motives. Why was the National Socialist leadership willing to continue to support their work? For years, many of the uranium scientists, together with allies in industry and the Armed Forces, had tirelessly stressed with considerable success the military importance of modern physics in general and of nuclear fission in particular. Some of the scientists did begin to downplay nuclear weapons in the last years of the war, during the desperate search for wonder weapons, but the various military and governmental officials had hardly forgotten their earlier lesson. The National Socialist state and Armed Forces were more than willing to encourage the uranium project—so long as it did not interfere with the war effort—because they recognized that such powerful new weapons would be very useful after the war.

✠   ✠   ✠

Purgatory (April 1945–1953) The unconditional German surrender in May 1945 was followed by the occupation of Germany by the four victorious powers: Britain, France, the Soviet Union, and the United States. This postwar period is very important for an understanding of the interaction between German science and National Socialism because the manner in which German scientists now dealt with Hitler's legacy reveals a great deal about how these scientists had perceived their work during the Third Reich. Although most scientists were happy that the war was finally over, they were ambivalent about what lay ahead. The Allies' announcement that they would strictly control scientific research and both denazify and demilitarize Germany threatened the scientists' future.

✠   ✠   ✠

Denazification The military effort against Germany had been portrayed during World War II as a struggle against the evil of National Socialism. But after the fall of the Third Reich, the victorious allies could only agree on their *intention* to purge public life entirely of National Socialist influence. From the very beginning the four powers' fundamentally distinct perceptions of the causes and supporters of National Socialism created grave differences with regard to the timing and scope of denazification.[695] In the Soviet zone denazification played an important role in the construction of a new social order based on the Soviet model. In the western zones denazification was essentially restricted to a comprehensive political purge of personnel and left the economic sphere basically untouched. Finally, whereas denazification was a pillar of American occupation policy, it was much less important for the more pragmatic British and the French.

The occupying forces initially ran the denazification themselves, often with catastrophic consequences for public administration and the economy. Mere membership in the NSDAP or an ancillary organization could be grounds for dismissal pending denazification, a policy which had the predictable effect of forcing solidarity in the face of this blanket threat—including among scientists. However, denazification was quickly turned over to the Germans themselves, both in order to save money and because only Germans were in a position to make the necessary differentiated judgments of conduct under National Socialism. Denazification was now recast as a judgment of personal responsibility, not mere membership in a political organization, and was transformed into a "Factory for Fellow Travelers."[696]

When the four powers decided to wrap up denazification by early 1948, only part of the German population had been investigated. Many entrepreneurs, bureaucrats, and professionals were temporarily exempted from the process in order to facilitate the reconstruction of Germany. Other individuals with money or influence managed to have their cases delayed or appealed. Denazification was thus effectively stopped at a point where most of the "little Nazis" had been through the process, but the big fish escaped relatively unscathed. In retrospect, denazification seems to have been doomed to failure. Without the agreement and cooperation of the Germans, a political purge like denazification could be administratively ordered from above, but not effectively carried out.[697] Of course, this statement holds just as well for the purge of German society by the National Socialists.

Since the western zones were on their way to becoming democracies, their politicians had to cater to the majority will, which was hardly enthusiastic about denazification or self-critical with regard to conduct during the Third Reich. The universities and research institutes were burdened with anti-democratic elements which long outlived National Socialism. This ideological baggage was a serious problem, for democracy can hardly work well when a large portion of those voting are essentially anti-democratic.

Perhaps the fundamental question is the *meaning* of denazification: did the allies intend to neutralize the threat of a National Socialist revival, or to punish previous conduct? In any case, scientists and in particular physicists were nothing special in this regard. Just like scientists had been subjected to the 1933 National

Socialist purge because they were part of the civil service, if they wanted an academic career after the war, then they had to endure the general denazification of the universities.

The denazification which began in 1945 was as much of a political purge as was the nazification that had started in 1933. No one asked in 1945 whether these scientists were good physicists or qualified teachers: they were judged by *political* criteria. Far fewer physicists were purged after 1945 than 1933. There was a severe shortage of physicists in Germany after World War II, so that pragmatism is one part of the explanation. This difference is also due in part to the fact that the dismissals and expulsions of 1933 were sometimes *racial* in nature, a criterion not employed after 1945. But this factor does not suffice to explain the stark contrast between 1933 and 1945.

Apologia can help illuminate this process. According to the usual postwar party line of the established German physics community, German physics remained apolitical during the Third Reich but had fallen behind American science because the National Socialists had ruined German science. However, there was a contradiction here, for the same scientists also asserted that the German physicists who were in place after the dust of denazification had settled were of high quality. Obviously if the National Socialists ruined physics, then some of the many physics professors who began their careers after 1933 and held positions after 1949 should be incompetent political appointees. Conversely, if the postwar physics community was of such high quality, then how could the Third Reich have ruined German physics?

In fact, when attrition due to aging and the postwar employment of physicists by the victorious powers are taken into account, the very small group of physicists purged after 1945 is practically equivalent to the equally small number of former adherents of *Deutsche Physik*. It is no surprise that denazification barely touched physics—it barely touched almost everything—but that in no way explains why only *Deutsche Physik* was purged. No other subset of German physicists, including former SS physicists, was punished so thoroughly and zealously.

A crucial portion of the new party line ran as follows: the physicists who had rejected *Deutsche Physik*, almost no matter what else they had done during the Third Reich, were now practically portrayed as resistance fighters; while the former supporters of *Deutsche Physik*, almost no matter what else they had done under Hitler, were branded "Nazis." But the latter were hardly the only physicists who had collaborated with National Socialism.

The political charges levied against the former followers of Lenard and Stark were usually accompanied by often unfair criticism of their scientific ability. In fact, although they were certainly not the best German physicists, they were also not all incompetents. Thus the final piece of postwar apologia fell into place. Whereas the competent and talented "real" physicists had resisted *Deutsche Physik* and thereby Hitler, only the followers of Lenard and Stark, who hardly deserved the name of scientist, had served National Socialism.

After the war the former followers of Lenard and Stark naturally tried to defend themselves when attacked and to avoid part or all of the punishment headed their way. They hoped to hold on to their positions and pensions, and to avoid fines or, in the most extreme cases, imprisonment. The occupying powers in turn were most interested in the utilitarian value of German physics, not in denazification. Just like influential actors in the National Socialist state had sided with modern physics because it promised to further their political and military goals, the occupying powers chose to back the same scientists because they might be able to help win the Cold War.

But why did the established physics community consciously create and consequently use Deutsche Physik as a scapegoat? Heisenberg, who had never joined a National Socialist organization and in postwar Germany enjoyed the status of a "victim of the Nazis," had a great deal of influence as the author of "whitewash certificates," written testimonials designed to help an individual pass unscathed through the process of denazification. Heisenberg, for instance, helped the convinced National Socialist physicist Pascual Jordan[698] and the SS physicist Johannes Juilfs[699] receive university appointments. In contrast, when Johannes Stark was tried for denazification, Heisenberg went out of his way to condemn his elderly colleague.[700]

By asserting that only *Deutsche Physik* had been politicized under National Socialism, the established physics community could kill several birds with one stone. First, they appeared to be participating wholeheartedly in the denazification of their profession, to be putting their own house in order. Second, they managed to avoid the purge or punishment of the overwhelming majority of their colleagues. Finally, by coupling scientific incompetence with service for the National Socialists, both of which they restricted to the *Deutsche Physik*, they tacitly asserted that their profession was inherently apolitical and a trustworthy servant worthy of generous support.

✠   ✠   ✠

Demilitarization When the occupying powers called for denazification, it was in the context of denazification *and* demilitarization. Physics and science had certainly been militarized during the Third Reich, indeed this transformation was an inevitable consequence of the strategy the established physics community had employed in order to defeat *Deutsche Physik*. However, German demilitarization proved just as ambiguous as denazification. Did the allies intend to neutralize German militarism or to punish previous militarism? To stop all German contributions to militarism or to demilitarize the German nation?

The demilitarization of German science was fundamentally and perhaps inevitably hypocritical.[701] Each of the four victorious powers hunted down German scientists and engineers as intellectual reparations. The Soviets called their researchers "specialists," a fitting name which underscored how the former allies perceived and treated their former enemies. The armorers of the National Socialists were now judged by what they could do for their new employers, not for what they had done for Hitler.

German specialists contributed significantly to the postwar science of all four victorious powers, although these countries have only grudgingly acknowledged that Germans worked for them, let alone that these specialists played an important role. Work for a foreign power obviously had its disadvantages, especially if it was coerced, but these researchers benefited as well. Their working conditions and compensation were relatively good, they could continue their work at a time when such research was often banned in Germany, and they did not have to go through denazification or justify their past political conduct.

Denazification and demilitarization had an important effect on German science, but not necessarily what had been intended. After World War II the victorious powers as well as the two new German states were in complete agreement with their scientists. If physics was useful, and what is more useful than powerful new weaponry, then physicists would be used. Physicists were seen first and foremost as tools, and tools do not need to be denazified or demilitarized. Physics in both the East and the West was materially rebuilt in order to serve one of the two sides in the Cold War. By the fifties, German physics in general was a solid and well-integrated, if subordinate part of the international scientific community.

Nazification and militarization had an unforeseen long-term effect on German science: it provided a push towards the "Big Science" so typical of the post–World War II period. Academic scientists were compelled to work in interdisciplinary research teams and closely with the government, the Armed Forces, and German industry. On the other, more negative side, a generation of physicists had been lost through the neglect and politicization of the education system as well as the terrible war. Science also suffered during the destructive chaos at the end of the war and immediate postwar period.

The overwhelming majority of scientists passed through de-nazification unscathed, but with the need to justify their previous work under Hitler. The denazification and demilitarization of German scientists and engineers had a profound effect on their self-image and postwar myths. Service for a victorious power— whether voluntary or not—retrospectively justified previous work for the National Socialists and facilitated apologia.[702] After all, how could a researcher's work during the Third Reich be criticized, when the Soviets or Americans wanted these same scientists and engineers to continue their work in the Soviet Union or the United States?

Did the Germans *try* to build atom bombs? If under *try* one understands the obviously necessary investments worth billions of dollars, the construction of huge factories, the development of suitable detonation devices, etc., then they *did not* try. But if under *try* one understands the manufacture of substances which were known to be potential nuclear explosives, and indeed the efforts to manufacture them as quickly and on the greatest scale possible without hindering the war effort, then they *did try*. The question perhaps most often asked, did the Germans *try to build an atom bomb, has no simple answer.*

9

The Crucible of Farm Hall

Why didn't Hitler get the bomb? Traditionally this question has been answered by scientists and historians alike in a black-or-white fashion. Either the team of German scientists were incompetent National Socialist collaborators or they had resisted Hitler by denying him nuclear weapons. Both claims are problematic. Once again, the truth lies somewhere in the middle.

One of the most controversial parts of the history of "Hitler's Bomb"[703] is the long-running debate over the mysterious and elusive "Operation Epsilon" recordings. These conversations, which have only recently been released, were recorded immediately after the war and without the knowledge of ten German scientists detained after the war at Farm Hall, an English country house near Cambridge.[704] General Leslie Groves' *Now It Can Be Told*, the immodest memoirs of the former head of the American atom bomb project, revealed in 1962 that the conversations of the Farm Hall scientists had been recorded, and that transcripts of these conversations existed.[705] But Groves provided only brief excerpts from the transcripts. In retrospect, the naturalized Ameri-

Farm Hall, 1945. (From the National Archives and Records Services.)

can physicist Samuel Goudsmit apparently used the Operation Epsilon report when writing his 1947 book *Alsos*.[706]

Samuel Goudsmit and the Alsos Mission came to Germany in the wake of the advancing Allied armies in order to determine and neutralize the threat of German nuclear weapons. When the investigation was finished, the Alsos Mission had seized or destroyed most of the material and scientific reports it found and arrested ten German scientists: Erich Bagge, Kurt Diebner, Walther Gerlach, Otto Hahn, Paul Harteck, Werner Heisenberg, Horst Korsching, Max von Laue, Carl Friedrich von Weizsäcker, and Karl Wirtz. They were brought to Farm Hall after brief stops in France and Belgium.

Since all but one of these scientists had been active in the German uranium project, they rightly assumed that they had been arrested because of their research. Ironically, they also falsely assumed that they were ahead of the Allies. Two concerns preoc-

Samuel Goudsmit, date unknown. (Courtesy of the AIP Emilio Segrè Visual Archives.)

cupied the guests: they were troubled by their inability to communicate with the families they had been forced to leave behind and they had no idea when or if they could go home.

In time, the Farm Hall detainees also confronted themselves with five fundamental questions:

1. Was I a "Nazi"?
2. Did we know *how* to make atom bombs?
3. Could Germany under National Socialism have produced nuclear weapons?
4. Did we *want* to make atom bombs?
5. What about our future?

Erich Bagge, 1945 at Farm Hall. (From the National Archives and Records Services.)

Here we will examine these questions and the answers these scientists reached in the context of the Third Reich and postwar Germany. Unless otherwise designated, all of the comments made by the Farm Hall detainees were private conversations, not statements to their jailers. Although the ten German scientists could have suspected that they were being monitored, it appears that they did not.

Kurt Diebner, 1945 at Farm Hall. (From the National Archives and Records Services.)

✠   ✠   ✠

Was I a "Nazi"? The first question to trouble the scientists was whether they bore personal responsibility for part or all of the excesses of National Socialism. In other words, who were the "Nazis" among them? Only Erich Bagge and Kurt Diebner had been members of the National Socialist Party, but only Otto Hahn, Werner Heisenberg, and Max von Laue had not joined some

Otto Hahn, 1945 at Farm Hall. (From the National Archives and Records Services.)

National Socialist organization.[707] Diebner, a civil servant in Army Ordnance, had held far more responsibility than his younger colleague Bagge, so that it was no surprise that he acted defensively at Farm Hall.

First, Diebner said that he only stayed in the Party because if Germany had won the war, then only NSDAP members would have been given good jobs. Next he argued that he had suffered under National Socialism. He had never voted for Hitler during

Walther Gerlach, 1945 at Farm Hall. (From the National Archives and Records Services.)

the Weimar Republic. In 1933 he became a Freemason in opposition to National Socialism. Once this information became known, Diebner claimed, he had experienced difficulties, at the university institute he was affiliated with and at Army Ordnance, where his promotion to civil servant was delayed. Furthermore, Diebner claimed that he had prevented the German looting of the physics institute in Copenhagen[708] and the arrest of Norwegian colleagues during the war, thereby tacitly coupling the responsibility he had as a Party member in Army Ordnance with the ability to restrain National Socialist excesses.[709]

Diebner's colleagues at Farm Hall were not so understanding. Otto Hahn pointedly remarked that being in the NSDAP had not done him any harm. When the scientists subsequently were considering drafting a written statement which would claim that their group had taken an "anti-Nazi" stance during the Third Reich, both Walther Gerlach—one of Diebner's few defenders—and Werner Heisenberg said that they could not conscientiously sign any such statement if Diebner had signed it as well. Diebner himself had no illusions about his future. He feared that when he returned to Germany, everybody would label him a Party member.[710]

For his part, Erich Bagge argued that he and the rest of the young assistants had been pressured into joining the University Stormtroopers, and that he had entered the NSDAP unknowingly. When someone asked his mother in the autumn of 1936 whether Bagge had wanted to join, she thought that it was a good thing and sent in his name. A few months later Bagge received his Party book which falsely said that he had been in the Party since 1 May 1935 and had sworn an oath to Adolf Hitler.[711]

Bagge generally was treated much more sympathetically by his colleagues than was Diebner. Heisenberg explained to a visiting English colleague and friend that Bagge had come from a proletarian family, which was one of the reasons why he joined the NSDAP, but that Bagge had never been a "fanatical Nazi." However, Gerlach rejected the suggestion that anyone had to join the Party, thereby stirring up considerable animosity. Once Gerlach had left the room, Bagge remarked that Gerlach had been protected from political attacks because he knew Göring personally and had a brother in the SS. Indeed Gerlach's jailers believed that he was particularly concerned to distance himself from National Socialism. Perhaps, they speculated, he had a guilty conscience.[712]

But there was more to being a "Nazi" than Party membership. The British wardens detected the lingering effect of National

Socialist ideology. Bagge expressed grave concern at the fact that Moroccan French soldiers had been billeted in his house. Bagge was not alone. When the detainees were lent a copy of *Life* magazine containing articles on the atom bomb and a number of photographs of scientists, von Weizsäcker remarked that of course they were mostly German, even though this statement was in fact untrue. The British commander reacted by reporting the conceit of the Germans who, with the possible exception of von Laue, still believed in the Master Race.[713]

Finally, the scientists expressed very different opinions about the worst excesses of the National Socialists. Bagge argued that if the Germans had put people in concentration camps during the war—he did not do it, knew nothing about it, and always condemned it when he heard about it—and if Hitler had ordered a few atrocities in concentration camps during the last few years of the conflict, then these excesses had occurred under the stress of war. In contrast, Karl Wirtz stated flatly that he and his countrymen had done unprecedented things. In Poland Jews were murdered. The SS also drove to a girls' school, Wirtz added, fetched out the top class and shot them simply because the Polish intelligentsia was to be wiped out. Just imagine, he asked his colleagues, if the Allies had arrived in Hechingen, the small town where Wirtz's institute had been evacuated during the last years of the war, driven to a girls' school and shot all the girls! That's what "we" Germans had done, he said.[714]

Perhaps the most interesting aspect of this moral question, who was a "Nazi"? is that this discussion practically vanished once these scientists heard the news of Hiroshima. Other questions now preoccupied their minds.

✠   ✠   ✠

Did We Know How to Make Atom Bombs? When Goudsmit (and the others who have subsequently taken up his arguments) asserted after the war that Heisenberg did not understand how an atom bomb worked,[715] there were three parts to his supposed lack of understanding: (1) Heisenberg had not realized that plutonium was fissionable material suitable for a nuclear explosive; (2) that nuclear weapons used fast-neutron chain reactions; and (3) only relatively small amounts of fissionable material were needed. Put these three together and you get Goudsmit's claim that the Germans in general and Heisenberg in particular mistook the nuclear reactor they were building for an atom bomb.

There is ample evidence that Heisenberg understood during the war that uranium 235 and plutonium were fissionable materials suitable for nuclear explosives and that such nuclear explosives used fast-neutron chain reactions.[716] The Farm Hall transcripts also corroborate Heisenberg's consistent understanding of these two areas.[717] All that was left was the matter of critical mass for a bomb.

Fortunately, a comprehensive February 1942 Army Ordnance report on the German uranium program includes the statement that the critical mass of a nuclear weapon lay between 10 and 100 kilograms of either uranium 235 or element 94.[718] There was no mention of who had made the estimate, and there was no reference to a scientific report which contained the calculation of the estimate. It seems most likely that Heisenberg would have been entrusted with this task, but he may have delegated the assignment, like he did many others.

Arguably it does not matter who made the estimate of critical mass or how it was made. German Army Ordnance decided in January or February of 1942 not to mount the industrial-scale effort which would have been needed to build nuclear weapons. The important question is: was the Army decision based on accurate information, comparable to that used in the United States? Or did German scientists mislead their military by exaggerating the difficulty of building the bomb?

In fact the German estimate of critical mass of 10 to 100 kilograms was comparable to the contemporary Allied estimate of 2 to 100. Thus the decision made by Army Ordnance was based on accurate information. The German scientists working on uranium neither withheld their figure for critical mass because of moral scruples nor did they provide an inaccurate estimate as the result of a gross scientific error. Instead the Army decision should be attributed to the differences in context between the Germans and the Allies: for example, how long each of the two sides assumed the war would last, the availability of raw materials and manpower, and the effect of the fighting on the war economy.[719]

The Operation Epsilon transcripts tell us what these scientists knew about nuclear weapons. On 6 August 1945 the detainees learned of the detonation of an American atom bomb.[720] At first they did not believe their English wardens, but after hearing the official announcement later in the evening they realized that the news was true. Hahn immediately asserted that the Allies must have managed to separate the isotopes of uranium, thus producing pure uranium 235, a nuclear explosive. But his colleague Paul Harteck, who used centrifuges during the war in an effort to achieve uranium isotope separation, reminded Hahn that another nuclear explosive, the transuranic element 93, could be manufactured in a nuclear reactor.[721] They did not yet know how the Allies had built their bomb.

This exchange also illustrates one reason why the brief excerpts from the Farm Hall recordings published by Groves have been misinterpreted. Even though all concerned had already demonstrated their knowledge of the fact that 93 decays within 2.3 days to a stable element 94 (plutonium), in their informal conversation the Germans usually used the term 93. The explanation for this apparent sloppiness in terminology may be traced back to the fact that during the war Kurt Starke, a young scientist working in Hahn's lab, had succeeded in separating out and analyzing 93, but though they were certain element 93 would produce 94, neither he nor his senior colleague had managed to produce plutonium.[722]

Heisenberg was one of the most skeptical scientists with regard to the Allied atom bomb. At first he did not believe a word of the report, but hastened to add that he could be wrong. Then he made a curious remark: it was perfectly possible that the Americans had ten tons of enriched uranium, but not ten tons of pure uranium 235.[723] Hahn immediately questioned Heisenberg's statement. During the war the physicist had told Hahn that only a

Carl Friedrich von Weizsäcker, 1945 at Farm Hall. (From the National Archives and Records Services.)

relatively small amount of uranium 235, 50 kilograms, was necessary; why was Heisenberg now saying that tons were needed?

Heisenberg responded by saying that for the moment he would rather not commit himself. He did say that if the bomb had been made with uranium 235, then the Germans should be able to work out exactly how it had been done. It just depended on the order of magnitude, whether it was done with 50, 500, or 5,000

Karl Wirtz, 1945 at Farm Hall. (From the National Archives and Records Services.)

kilograms of fissionable material. He went on to say that the Germans could at least assume that the Americans had some method of separating isotopes, even if the scientists at Farm Hall did not know what that method was.[724] Heisenberg did return to this question of critical mass before he left Farm Hall.

Carl Friedrich von Weizsäcker and Karl Wirtz debated whether the Americans had used plutonium for their nuclear explosive.[725] Von Weizsäcker in fact had brought the potential use of transuranic elements as nuclear explosives to the attention of Army Ordnance in 1940.[726] Wirtz was skeptical, but not because he was ignorant of what needed to be done to manufacture plutonium. Von Weizsäcker agreed. The Allied scientists who had captured them in Germany had showed much more interest in isotope separation, so that von Weizsäcker assumed that they had used the same method.[727]

The official announcement at 9:00 in the evening stunned the Germans because they now realized that the news was genuine. Harteck asserted that the Allies had managed to make a bomb either by using electromagnetic uranium isotope separation on a large scale—and of course the Americans did use this process along with other methods—or some photochemical isotope separation process.[728] Harteck's suggestion illustrates another reason why the Farm Hall transcripts have been misinterpreted. Although these scientists were aware that transuranic fissionable material could be manufactured in a nuclear reactor, most of them now assumed that the Americans probably used isotope separation to make uranium 235, and not a nuclear reactor to make transuranics, in order to make their nuclear explosives.

Thus Hahn remarked that the Allies seemed to have made a nuclear explosive without first perfecting the nuclear reactor.[729] This assumption was accepted by many of his colleagues as well, apparently because it allowed them to hold out hope (for at least a little while longer) that they had outperformed their British and American competitors in at least one area. Considering the newspaper accounts of the enormous scale and cost of the Allied effort, Harteck speculated that they must have used a huge number of mass spectrographs, since if they had had a better method, then it would not have cost so much. Even though Horst Korsching and Wirtz, both younger physicists with experience in isotope separation research, doubted that spectrographs had been used, Heisenberg and other senior scientists accepted Harteck's theory. This suggestion was plausible because the Germans knew that this

Werner Heisenberg, 1945 at Farm Hall. (From the National Archives and Records Services.)

technology was both available and could produce pure uranium 235.[730]

When they were alone, Hahn pressed Heisenberg again on the actual size of the atom bomb. If the Allies had set up a hundred thousand mass spectrographs, Heisenberg said, then they could produce 30 kilograms a year of uranium 235. Hahn responded by asking whether the Americans would need as much as that for a bomb? Heisenberg's answer to Hahn's question is illuminating: yes, he thought that the Allies would certainly need that much fissionable material, but quite honestly, he told Hahn, he had never worked it out.[731]

Hahn then asked how the bomb exploded? Heisenberg first responded with a rough argument using the mean free path of a fast neutron in uranium 235 to get an improbably large estimate of the radius of critical mass: 54 centimeters, which would mean a ton of 235. But he immediately went on to say that the Allies could have done it with less, perhaps a quarter of that quantity, by using a fast neutron reflector or tamping around the critical mass. In 1943 the young German physicist Karl-Heinz Höcker had worked out the theory for a nuclear reactor using a lattice of uranium spheres, calculating both the diffusion of fission neutrons in a spherical mass of fissionable material and the probability that the surrounding spherical layer of moderator would reflect neutrons back into the sphere. Moreover, it is known that Heisenberg followed Höcker's work closely.[732]

Hahn also asked Heisenberg how the Americans could have taken such a large bomb in an aircraft and be certain that it would explode at the right time? His physicist colleague replied that the bomb could be made in two halves, each of which would be smaller than the critical mass. The two halves would then be joined together to ignite the chain reaction.[733]

In response to a subsequent question from Gerlach, Heisenberg also speculated that perhaps the nuclear explosive was merely enriched uranium, some mixture of the isotopes 235 and 238.[734] Heisenberg was certainly aware that pure 235 would be better than any mixture, and in 1939 he had told Army Ordnance that pure 235 was needed for such an explosive.[735] He was apparently so skeptical at Farm Hall that the Allies could have succeeded in total uranium isotope separation that he was willing to consider the possibility of using enriched uranium in an atom bomb—a strategy which would not have worked.

On 8 August 1945 the detainees read in the newspapers that the Americans had used "pluto" in a bomb, and there was imme-

Paul Harteck, 1945 at Farm Hall. (From the National Archives and Records Services.)

diate speculation as to whether this new element was element 94. This newspaper account provoked another illuminating remark from Heisenberg. The Germans had not even attempted to research fast neutron reactions in 94 because they did not have this element, and saw no prospect of being able to obtain it.[736]

The following day the newspapers mentioned that the atom bomb weighed 200 kilograms, prompting a conversation between Harteck and Heisenberg. Harteck asked whether this was the true weight of the bomb or whether the Americans were merely trying to bluff the Russians. This latest piece of information worried Heisenberg, because it suggested that his estimate of critical mass was too large. He decided to take another look at the problem.

An important part of his previous calculations was the multiplication factor of fission neutrons: how many neutrons would each nuclear fission release? Heisenberg had been using a conservative multiplication factor, 1.1, the value they had observed during their own uranium machine experiments. When Heisenberg substituted a factor of 3, he found that the radius of the critical mass was comparable to the mean free path, roughly 4 centimeters, which made the critical mass considerably smaller.[737]

Harteck and Heisenberg then reconsidered the possibility of using 94 as a nuclear explosive. Heisenberg pointed out that the use of 94 would mean that the American uranium machine had been running since 1942. Moreover, the chemical separation of 94 from uranium would be fantastically difficult. Harteck, an accomplished physical chemist, agreed with Heisenberg that it was highly improbable that the Allies had succeeded with 94.

The detained scientists continued to discuss how their Allied colleagues had managed to manufacture an atom bomb. Eventually Heisenberg was asked to give a lecture on the subject. Such talks were common at Farm Hall. The detainees entertained themselves and kept busy by holding an informal series of scientific lectures. The presentation, which was punctuated by questions and lively debate, took place on 14 August 1945. By this time, Heisenberg asserted that they (in other words, he) understood very well how the atom bomb worked.[738]

Heisenberg now assumed that 2 to 2.5 neutrons were released per fission. He used a diffusion equation for neutron density, assumed that there was a neutron reflector surrounding the fissionable material, and calculated a critical radius of between 6.2 and 13.7 centimeters for the atom bomb. Heisenberg was still dissatisfied, because the newspaper article claimed that the whole explosive mass only weighed 4 kilograms, but the sphere with a 6.2 centimeter radius would weigh 16.

In his Farm Hall lecture Heisenberg went on to discuss a possible detonation mechanism for the bomb. Two hemispheres, each slightly smaller than the critical mass, would be placed in an iron cylinder, actually a gun barrel, such that one hemisphere would be shot at the other. Indeed the Hiroshima bomb did use such an arrangement. Finally, Heisenberg speculated on the effect of the nuclear blast. The first 10 meters of air surrounding the bomb would be brought to a white heat. The surface of the uranium sphere would radiate about 2,000 times brighter than the sun. It would be interesting, he added, to know whether the pressure of this visible radiation could knock down objects.

Four days later, one of the English officers showed the detained scientists the British White Paper on the atom bomb, an official publication which effectively cut off all further speculation by the Germans on the technical aspects of Allied nuclear weapons.[739] Apparently the wardens at Farm Hall were now confident that the Germans had revealed everything they knew about nuclear weapons. Heisenberg now noted that the physics of it was actually very simple. It was an industrial problem and it would never have been possible for Germany to do anything on that scale.[740] Thereafter the Germans spent their time worrying about their future and trying to get back home.

The transcripts of Operation Epsilon also provide additional evidence for dismissing the postwar claims by Heisenberg and others that Bothe's "mistake"—he had measured the diffusion length of thermal neutrons in carbon—slowed down the German effort by diverting their efforts away from the use of graphite as neutron moderator towards heavy water.[741] There is absolutely no mention of graphite as a moderator in the Farm Hall transcripts. Only after Heisenberg and others had read the official American publication[742] on the atom bomb, and thereby learned that the Americans had used graphite, did they begin to use Bothe as a scapegoat, the one German scientist whose error had handicapped their efforts. In fact Army Ordnance had considered using graphite as a moderator, but chose heavy water because it appeared less expensive.[743]

The postwar accounts by Groves and Goudsmit of Farm Hall are sometimes distorted. Statements from the Operation Epsilon transcripts are often taken out of context and other remarks, which would make clear what these Germans did and did not know, are passed over in silence. Goudsmit describes how the detainees debated what the "plutonium" mentioned in the newspaper accounts meant, but does not also say that the Germans had been discussing the transuranic elements 93 and 94 and their properties throughout their captivity.

The question for Heisenberg, Hahn, Harteck, and the rest of their colleagues was whether the Allied plutonium was what they knew as 94, and subsequently the Germans reached a consensus that it was. Similarly, Goudsmit tells us that Heisenberg and the others speculated whether perhaps the Allies had used the radioactive element protactinium as an explosive, but without making clear that this speculation was in the context of either uranium 235, or plutonium, or protactinium as an explosive.[744]

Groves is sometimes unfair in his handling of Heisenberg. He faithfully reproduces Heisenberg's statement admitting both ignorance of how the Allies succeeded and the disgrace he felt that they did not know how their British and American colleagues had done it. But Groves does not tell the reader that Heisenberg's statement is preceded by a long and surprisingly accurate speculation on exactly how the Allied atom bomb worked.[745] Finally, Goudsmit makes several claims that are simply wrong and for which there is no supporting evidence in the Farm Hall transcripts: (1) that the Germans believed that the Americans had dropped a complete nuclear reactor on Hiroshima; (2) that at first the Germans had not understood that the plutonium used as an explosive is produced in the reactor; and in short (3) that the Germans had failed to realize that there is a difference between a reactor and an atom bomb.

But the most controversial technical aspect of the Farm Hall recordings has always been Werner Heisenberg's apparently confused conception of an atom bomb. His understanding that fast neutron chain reactions in pure uranium 235 and plutonium constituted nuclear explosives had been demonstrated during the war and is reinforced in the Operation Epsilon report as well. The one unclear point is critical mass of the weapon: how much was needed for the bomb to go off?

In contrast to Groves and Goudsmit, both R. V. Jones' and Charles Frank's accounts from memory of the Farm Hall recordings were quite accurate. The British scientist Jones remembered Heisenberg's first "back-of-the-envelope calculation" for critical mass, whereas his countryman Frank in turn remembered Heisenberg's subsequent sophisticated calculation using a "rather polished version of diffusion-and-multiplication theory."[746]

During the war Heisenberg most probably made a rough estimate which was comparable to contemporary Allied estimates, but more importantly was good enough for German Army Ordnance to decide not to attempt the industrial-scale production of nuclear weapons. At the time the German researchers had been unable to separate out uranium 235 or to sustain a chain reaction in a uranium machine. Even this relatively small critical mass must have appeared out of reach until after the war.[747] Heisenberg himself admitted at Farm Hall that he had never made a more precise calculation of critical mass, not because he was incapable of it, but because there was no point. R. V. Jones has even speculated that Heisenberg made an accurate calculation in 1942, but had forgotten it by the summer of 1945.[748]

Groves' and Goudsmit's assessments were probably colored by their desire to "prove" that the Germans had been incompetent and thus saw in these transcripts what they wanted to see. But they also called Heisenberg's scientific abilities into question for a specific reason: to explain why the Germans did not make an atom bomb. If the Farm Hall recordings make anything clear, it is that Heisenberg's temporary confusion with regard to critical mass had nothing to do with the scale, tempo, or success of the German efforts to harness the military applications of nuclear fission. Anyone who wants to know why the world never saw National Socialist nuclear weapons will have to look far beyond Farm Hall.

✠   ✠   ✠

*Could Germany under National Socialism Have
Produced Nuclear Weapons?* It is important to separate the
question, did the German scientists *know how* to make atom bombs,
from two other questions: (1) *could* the Third Reich have manufactured nuclear weapons before the end of the war; and (2) did these
scientists *want* to make atom bombs for the National Socialists? The
press reports of the attack on Hiroshima and Nagasaki touched off
a heterogeneous reaction among the scientists, a reaction which
moreover changed over time.

Karl Wirtz was one of the few detainees to simply and flatly
say that he was glad that they did not have the atom bomb.[749] Otto
Hahn's reaction was similarly unambiguous: he would have sabotaged the war effort if he had been in a position to do so. When he
was privately told the news before the rest of his colleagues, it
shattered his composure. He told his warden that he had originally
contemplated suicide when he realized the destructive potential of
his discovery of nuclear fission, and that he now felt personally
responsible for the deaths in Hiroshima. Several alcoholic drinks
were required to calm Hahn down sufficiently to let him rejoin his
colleagues.[750]

The reaction of Walther Gerlach, who had been in charge of
the uranium research during the last eighteen months of the war,
was quite different. He went up to his bedroom and began to cry,
despite the efforts of Paul Harteck and Max von Laue to comfort
him. Gerlach's British captors saw him acting as a defeated general
and contemplating suicide. Hahn subsequently asked him why he
was so upset. Was it because Germany did not make an atom bomb
or because the Americans could do it better than the Germans?[751]

Gerlach insisted that he was not in favor of inhuman weapons like the atom bomb. In fact he had been afraid of it and had not
believed that the bomb could be made so quickly. But he was
depressed because the Americans had demonstrated their scientific superiority. He realized during the last years of the war that
the bomb would eventually be developed, and was determined to

xploit the potential of uranium for Germany's future. Thus he told
Colonel Geist, Minister of Armaments Albert Speer's right-hand
man, and Fritz Sauckel, Plenipotentiary for Labor Development,
that he who could threaten the use of the bomb could achieve
anything.[752]

Heisenberg later explained to Hahn that Gerlach was taking
the news so badly because he was the only one of the Farm Hall
scientists who had really wanted a German victory. Although
Gerlach had known and disapproved of the crimes of the "Nazis,"
he felt that he was working for Germany. Hahn replied that he,
too, loved his country, and that as strange as it might seem, that
was why he had hoped for her defeat.[753] Gerlach himself went
further, tacitly criticizing the Allies by arguing that, if Germany
had had a weapon which would have won the war, then Germany
would have been in the right and the others in the wrong. Moreover, conditions in Germany were not now better than they would
have been after a Hitler victory.[754]

Gerlach was not the only one to criticize the Allies. Von
Weizsäcker called the American atom bomb attack on Japan mad-
ness. Heisenberg objected that one could equally say that using
nuclear weapons had been the quickest way to end the war,
whereupon Hahn added that that thought was what consoled
him.[755] Wirtz was horrified by Hiroshima and argued that it was
characteristic that the Germans had discovered nuclear fission but
the Americans were the ones who used it.[756]

When the news of Hiroshima began to settle in, several of the
scientists began to argue that they could not have made atom
bombs. Von Weizsäcker pointed out that, at the rate they had been
going, they could not have succeeded during the war. Even the
scientists involved with the research had said that it could not be
done before the end of the conflict.[757] Although Bagge rejected von
Weizsäcker's comment at the time, he subsequently admitted that
none of the scientists had forcefully pushed the project.[758]

Heisenberg put this question into the context of science policy during the Third Reich. In the spring of 1942, when the fate of
the uranium research was being decided, he would not have had

the moral courage to recommend that 120,000 men be employed–like in America—to move from research to development on th industrial scale. The entire German uranium project involved a most a few hundred workers. The relationship between the scien tist and the state under National Socialism, Heisenberg explained was at fault. Although he argued that he and his colleagues wer not 100% eager to make atom bombs, the scientists were so littl trusted by the state that it would have been difficult to accomplis even if they had wanted to do it.[759]

Kurt Diebner, who had been responsible for much of th administration of the uranium project, agreed, stating that th officials had been interested only in immediate results and did no want to pursue a long-term policy like the Americans had obvi ously done.[760] Harteck first argued that they might have succeeded if the authorities had been willing to sacrifice everything toward that goal, but upon reflection said that the Germans never coul have made a bomb, but certainly could have created a workin nuclear reactor. He was very sorry that they had failed to achiev the latter, no doubt because of the national and professional pres tige it might have meant.[761]

Von Weizsäcker also speculated at first that, if they ha gotten off to a better start, then the Germans might have ha nuclear weapons by the winter of 1944–1945. Wirtz pointedl replied that then Germany would have obliterated London, bu would still not have conquered the world, and then Allied aton bombs would have fallen on Germany. Von Weizsäcker agree that it would have been a much greater tragedy for the world i Germany had had the atom bomb.[762]

✠   ✠   ✠

Did We Want to Build Atom Bombs? The real contro versy surrounding the Farm Hall recordings has not revolve around whether these ten German scientists *could* have made aton bombs, rather whether they *would* have. The transcripts from Farn Hall demonstrate that von Weizsäcker did indeed eventually ar gue that, because they had not wanted to make nuclear weapons they did not. But these arguments were hardly a simple cover-up, rather a concerted attempt to persuade himself and his colleagues to revise their own memories in order to put a better face on an increasingly problematic past.

Von Weizsäcker began this reinterpretation by stating his belief that they had not made an atom bomb because all the physicists did not want to do it on principle. If they had all wanted Germany to win the war, then they would have succeeded. Hahn immediately rejected this suggestion,[763] and later Bagge privately said that it was absurd for von Weizsäcker to say that he had not wanted the thing to succeed. That might have been true in his case, Bagge allowed, but not for all of them.[764]

Von Weizsäcker's next step was to argue that, even if the German scientists had gotten all the support that they had wanted, it was by no means certain that they would have gotten as far as the Americans and British did. After all, the German physicists were all convinced that the thing could not be completed during this war. Heisenberg interjected that von Weizsäcker's interpretation was not quite right. Heisenberg had been absolutely convinced of the possibility of making a nuclear reactor, but never thought that the Germans would be able to make a bomb. Moreover, he admitted that at the bottom of his heart he was glad that only a reactor and not a bomb appeared possible. Here Heisenberg was being disingenuous. He was well aware that an operating nuclear reactor was perhaps the most important step towards making nuclear weapons.

Von Weizsäcker then pushed the point, arguing that if Heisenberg had wanted to make a bomb, then he would have concentrated more on isotope separation and less on a nuclear reactor. Otto Hahn left the room at this point, perhaps because he did not want to hear any more. Von Weizsäcker went on to argue again that they should admit that they did not want to succeed. Even if they had put the same effort into it as the Americans and had wanted it as badly, the Allied aerial bombardment of German factories would have doomed their efforts.[765]

This question, whether these scientists had *wanted* to succeed, was couched more and more in terms of moral principles. Heisenberg argued that, if the German scientists had been in the same moral position as the Americans, who felt that Hitler had to be defeated at all cost, then they might have succeeded. But Heisenberg and his colleagues had considered Hitler a criminal.[766] Indeed earlier at Farm Hall, when Heisenberg first learned of the agreement reached at the Potsdam Conference and the probable cession of German territory to Poland, he remarked that it would have been infinitely worse if Germany had won the war.[767]

But Heisenberg was clearly changing his mind with regard to his own past intentions. In a subsequent conversation with Hahn they both agreed that they had never wanted to work on a bomb and had been pleased when it was decided to concentrate everything on creating a nuclear reactor.[768] In fact no such decision was ever taken. Rather than dictating to the researchers that they would henceforth work on a reactor and not a bomb, Army Ordnance merely decided not to boost the research up to the industrial level.

This minor distortion of the historical record is important, for it forms a basic part of the postwar myths surrounding Hitler's bomb.[769] Still later, after Heisenberg had seen the British White Paper and thus knew a great deal about how the atom bomb had been achieved, he stated flatly in a conversation with his old friend and British colleague Blackett, who was visiting Farm Hall, that the Germans had been interested in a kind of machine, but not a bomb.[770]

But the most striking comment made in Farm Hall came from von Weizsäcker, who said that history would record that the Americans and English made a bomb, and at the same time the Germans, under the Hitler regime, produced a workable nuclear reactor. In other words, the peaceful development of the uranium machine was made in Germany under the Hitler Regime, whereas the Americans and the English developed this ghastly weapon of war.[771] The author Robert Jungk interviewed von Weizsäcker in

nuclear fission instead of serving his country as a soldier? He took up the work with enthusiasm, energy, and success. In two articles, finished respectively in December 1939 and February 1940, he worked out the theoretical foundations for nuclear energy and nuclear weapons and immediately passed them on to Army Ordnance.

5. How did Heisenberg react when it became clear to him in the autumn of 1941 that his colleague and former teacher Bohr was threatened in occupied Denmark, and separately, that in principle nothing stood in the way of nuclear weapons? He accepted an invitation to give a talk at a German astrophysics conference at the Copenhagen German Cultural Institute, a center for the cultural and scientific collaboration between native scientists and National Socialism. While he attended the Copenhagen conference, he also visited Bohr and told him: (1) Hitler would win the war; (2) nuclear weapons were possible; (3) the Germans were working on them; and (4) he, Heisenberg, had mixed feelings about it. Moreover, together with Carl Friedrich von Weizsäcker, Heisenberg advised Bohr to collaborate with the Germans and in particular with the German Cultural Institute.

6. How did Heisenberg react in February 1942 when Army Ordnance decided that nuclear weapons were not relevant to the war effort and that the uranium project would be transferred to the Reich Research Council in the Ministry of Education, decisions which threatened the financing and support of the project and thereby clearly endangered the security of the individual scientists? He lectured in February 1942 before leading figures in the National Socialist party and armed forces on the theoretical foundation of nuclear fission. This popular lecture made crystal clear both the military significance of nuclear fission in general and of Heisenberg's own work in particular, including the remark that nuclear explosives would have an "unimaginable effect." After a few weeks this information even landed on Josef Goebbels' desk.

7. How did Heisenberg react in 1943, 1944, and 1945 to the ever-deteriorating state of the war? Together with his other colleagues in the uranium project, he worked harder and harder,

desperately attempting to reach the now relatively modest research goals before the end of the war: to build a nuclear reactor which could sustain a nuclear fission chain reaction for a modest period of time; and to manufacture tiny amounts of pure uranium 235, that is, to create tiny amounts of a nuclear explosive.

8. How did Heisenberg react to the end of the war, when his Allied colleagues arrested him? Heisenberg made a distinct impression on them as an "anti-Nazi" and German nationalist.

9. When Heisenberg was interned in Farm Hall in England and heard the radio news of the bombing of Hiroshima, how did he react? He admitted only grudgingly that he had never made the calculations necessary for an atom bomb because he had believed that they would not be able to create them before the end of the war. Subsequently, he worked so intensely on this problem that after a few days he could explain to his colleagues in Farm Hall how the Allies had done it.

Yet what is most important is that over the next few weeks and with the strong encouragement of Carl Friedrich von Weizsäcker, Heisenberg began to change his opinion gradually and step-by-step. He said that he had not believed that these weapons were possible, and in his heart he had been glad. At the end, he said that he and his colleagues had not wanted to build nuclear weapons for Hitler and that these moral scruples were the reason why the "Nazis" did not get them.

10. How did Heisenberg react after World War II when his American colleague Samuel Goudsmit polemically attacked him? Goudsmit claimed that the Germans had not been in a position to build nuclear weapons because they had made crude, simple scientific errors. However, if they would have been in a position to do it, then they would have done what was necessary in order that Germany not lose the war. Heisenberg answered with an apologetic thesis co-authored by von Weizsäcker. The Germans had not been in a position to build nuclear weapons; but if they would have been in such a position, then they would not have done it. They would have done whatever was necessary in order that these horrible weapons not fall into Hitler's hands.

11. How did Heisenberg react to the denazification of the German physics community after World War II? He sharply criticized the former adherents of *Deutsche Physik* like Johannes Stark, but in contrast wrote "whitewash certificates" for those individuals who had worked against *Deutsche Physik*, almost no matter what else these persons had done during the Third Reich. Thus Heisenberg helped rehabilitate the SS-physicist Johannes Juilfs and the convinced National Socialist and physicist Pascual Jordan.

12. How did Heisenberg react to Robert Jungk's 1956 book, *Brighter Than a Thousand Suns*, which propagated the apologetic thesis and claimed that only a conspiracy around Heisenberg and von Weizsäcker had saved the world from National Socialist nuclear weapons? In 1957 Heisenberg explicitly corroborated the conspiracy theory in his private correspondence with Jungk, although in public he always restricted himself to hints and ambiguous remarks which tacitly strengthened the conspiracy theory.

13. How did Heisenberg react when the nuclear politics of Defense Minister Franz Josef Strauss and Chancellor Konrad Adenauer threatened public support for nuclear research and nuclear energy in the Federal German Republic, especially because this policy awakened the specter of "Hitler's bomb?" Together with von Weizsäcker and other colleagues like Otto Hahn, Max von Laue, and Karl Wirtz, he sent an open letter to Adenauer which made it clear that they would have nothing to do with the research, development, or stationing of nuclear weapons in Germany.

14. How did Heisenberg react in his later years when faced with his own mortality? In his 1969 memoirs, *Physics and Beyond*, he clearly implied that the conspiracy theory was true. In a 1970 private letter he claimed the conspiracy theory more explicitly than ever before. Together with Hahn and von Laue, Heisenberg had supposedly falsified the mathematical calculations in order to deny nuclear weapons to Hitler. This claim was not only false, it is tragically absurd.

Heisenberg may have resisted Hitler, in his own mind. Heisenberg's behavior was not so different from most of his colleagues in Germany, the United States, or the Soviet Union who worked on nuclear fission. Almost all of them cooperated with their governments under very different conditions, either out of conviction, ambition, or fear. There was an important difference, but that lay with the political, ideological, and moral nature of the regime, not the scientists.

But the main point here is not to condemn Heisenberg's conduct under National Socialism, rather to criticize activity by him and so many others since the end of the Third Reich. Why were myths and legends of active resistance against Hitler created and propagated after the war? Obviously because something is being repressed. Scientific work, exactly like any other occupation, can be politicized. Scientists in general are morally neither superior nor inferior to the general public. Finally, sometimes—for example under National Socialism during World War II—there are neither simple answers nor simple questions.

11

Conclusion
The Scientist as Fellow Traveler

This book began with a distinction between "Nazi" scientists and the scientists who served National Socialism. Yes, some scientists enthusiastically embraced Hitler's movement, but they are only part of the story. The German scientific community and most of its members entered into a Faustian pact with National Socialism, trading financial and material support, official recognition, and the illusion of professional independence for conscious or unconscious support of National Socialist policies culminating in war, the rape of Europe, and genocide.

Ironically, the initial ideological attacks on German science served to drive scientists into the arms of military, industrial, and political allies and thereby into an ever closer collaboration with Hitler's movement. This relationship, in contrast to the discredited and discarded politicized scientific movements like *Deutsche Physik*, was based not on ideology, rather utility. Johannes Stark was one of Hitler's first and most loyal followers, but although he was honored as an Old Fighter by the National Socialist movement, by the end of the Third Reich his party comrades had rejected him as a scientist and science policy maker in favor of "white Jews" like Werner Heisenberg.

The fact that German science collaborated with the National Socialists also explains the persistent and pernicious myth of scientific resistance to Hitler. If decent, moral men like Otto Hahn, Werner Heisenberg, Max Planck, Max von Laue, and Carl Friedrich von Weizsäcker stayed in Germany and worked within the National Socialist system, then, the story goes, they must have been passively or actively resisting Adolf Hitler. But the popular black-and-white picture of the Third Reich as a place where it was easy to see right and wrong, full of villains and heroes, is a simplistic caricature. It only appears that simple with hindsight. The history of the Prussian Academy and Heisenberg's guest lectures provide a more realistic picture of the Third Reich. National Socialism often pulled scientists very slowly and gently into its terrible grasp.

The ambiguous and ambivalent nature of National Socialism often makes it difficult to judge German scientists who stayed and worked for their government. But there is no such problem when judging the postwar apologia. After the war there was no totalitarian state or secret police to fear; after the war these scientists could speak their minds and tell the truth. The recorded conversations from Farm Hall demonstrate how these scientists reacted to the hard questions that followed the defeat of National Socialism. They spawned self-serving and self-deluding myths and legends which in time took on a life of their own. The unending debate over "Hitler's bomb" is only the most prominent example.

Finally, one question remains. Was science under National Socialism a special case? This question can only be answered through comparison with scientists in other countries or at other times, but such analysis is problematic. When historians begin to compare the Third Reich or the Holocaust with other regimes and genocides, some critics feel that the unique evil of National Socialism is being trivialized—and in some cases, for example the recent *Historikerstreit* (historians' dispute), they would be right.[852] But it must be possible to both respect the unique, terrible nature of National Socialism and compare it with other periods in history. The fact remains that French scientists served the Committee of Public Safety during the Reign of Terror,[853] Soviet scientists served Stalin,[854] and—admittedly a much less extreme case—most American scientists acquiesced in the excesses of McCarthyism.

Science under National Socialism remains so controversial, fascinating, and disturbing because of a *taboo*: the scientist as *fellow traveler*. The overwhelming majority of German scientists neither resisted Hitler, nor embraced National Socialism. Instead, they were apolitical professionals, doing what scientists do. Scientists often like to celebrate the "apolitical" nature of science. But one consequence of such science is taboo. It is precisely this apolitical nature of scientific research that allows good scientists to do good science, no matter who their employer or patron is or where this research may lead. Moreover, as far as science in the Third Reich was concerned, fellow travelers were often far more dangerous than either Hitler's true believers or his bitter opponents.

Abbreviations

AA	*Auswärtiges Amt* (German Foreign Office)
AdW	*Akademiearchiv der Berlin-Brandenburgischen Akademie der Wissenschaften* (Academy Archives of the Berlin-Brandenburg Academy of Sciences), Berlin
AIP	American Institute of Physics, College Park, Maryland
BAK	*Bundesarchiv* (Federal Archives), Koblenz
BAP	*Bundesarchiv* (Federal Archives), Potsdam
BDC	Berlin Document Center
DAAD	*Deutsche Akademische Austauschdienst* (German Academic Exchange Service)
DFG	*Deutsche Forschungsgemeinschaft* (German Research Foundation)
DMM	*Deutsches Museum* (German Museum), Munich
EBK	Erich Bagge Papers, Kiel

GCI   *Deutsches Wissenschaftliches Institut* (German Cultural Institute)

GKT   Gerald Kuiper Papers, University of Arizona, Tucson

HSA   *Hamburger Staatsarchiv* (Hamburg State Archive)

HUB   *Archiven der Humboldt Universität* (Archives of the Humboldt University), Berlin

IZG   *Institut für Zeitgeschichte* (Institute of Contemporary History), Munich

KWG   *Kaiser-Wilhelm Gesellschaft* (Kaiser Wilhelm Society)

LPG   Ludwig Prandtl Papers, Göttingen

MPI   *Max Planck Institut für Physik und Astrophysik* (Max Planck Institute for Physics and Astrophysics), Munich

MPGB   *Archiven der Max Planck Gesellschaft* (Max Planck Society Archives), Berlin

NAARS   National Archives and Records Services, Washington, D.C.

NBC   Niels Bohr General Correspondence, Niels Bohr Institute, Copenhagen

NG   *Notgemeinschaft der Deutschen Wissenschaft* (Emergency Foundation for German Science)

NRD   *Nachgeordnete Reichsdienststellen* (Subordinate Reich Agencies)

NSDAP   *Nationalsozialistische Deutsche Arbeiterpartei* (National Socialist German Workers Party)

NSLB   *Nationalsozialistische Lehrerbund* (National Socialist Teachers League)

OPG   *Oberstes Parteigericht* (Highest Party Court)

PTR   *Physikalisch-Technische Reichsanstalt* (Imperial Physical-Technical Institute)

REM   *Reichserziehungsministerium* (Reich Ministry of Education)

RIM	*Reichsinnenministerium* (Reich Ministry of the Interior)
RUB	*Rektor der Universität Berlin* (Rektor of Berlin University)
RUL	*Rektor der Universität Leipzig* (Rektor of Leipzig University)
SA	*Sturmabteilung* (Stormtroopers)
SBB	*Staatsbibliothek, Preussischer Kulturbesitz* (State Prussian Library), Berlin
SGA	*Sitzung der Gesamt-Akademie* (Meetings of the Full Academy)
SMV	*Sächsisches Ministerium für Volksbildung* (Saxon Ministry of Education)
SPM	*Sitzung der physikalisch-mathematischen Klasse* (Meetings of the Scientific Class)
SS	*Schutzstaffein* (Security Squad)

Endnotes

1. Renneberg and Walker, (1993).
2. My approach, using a spectrum of "shades of gray" to examine science under National Socialism, has been misunderstood by some critics; for example, the historian Alan Beyerchen has recently characterized my work as "blame assessment," "the identification not of the criminals but collaborators," and the "meting out of justice"; see Beyerchen, (1992), 632–4.
3. Hoffmann, (1982), 93.
4. Hoffmann, (1982), 91.
5. Kleinert, in Olff-Nathan (1993), 151.
6. Forman, "Helmholtz."
7. Kleinert, (1979), 501.
8. Kleinert, (1979), 501.
9. Kleinert, (1978).
10. Beyerchen, (1977), 92.
11. Kleinert (1979), 502–7; for Lenard, also see Beyerchen, (1977), 79–102.
12. Kleinert, in Albrecht (1993).
13. Hentschel, (1990), 134.
14. Hentschel, (1990), 133.

15. Kleinert (1979), 502–7.
16. Beyerchen, (1977), 92–3.
17. Hentschel, *Annals*, (1992).
18. Glaser to Stark (24 Sep 1920) Nachlass Stark, Mappe Glaser, Handschriftenabteilung Staatsbibliothek Berlin (henceforth SBB); for Nauheim, see Forman, "Nauheim.".
19. Glaser, (1920), 33.
20. Beyerchen, (1977), 111–12; Hoffmann, (1990), 44; Kleinert in Olff-Nathan (1993), 151–54.
21. Beyerchen, (1977), 104.
22. Kleinert (1979), 509.
23. Laue, (1923).
24. Beyerchen, (1977), 106.
25. Beyerchen, (1977), 122.
26. Spielvogel, (1992), 37–8.
27. Stark to the Oberstes Parteigericht (henceforth OPG) (7 Nov 1937) Stark, Berlin Document Center (henceforth BDC).
28. Beyerchen, (1977), 96.
29. Beyerchen, (1977), 4.
30. Information in the BDC.
31. Hoffmann, (1982), 95–6.
32. Stark to the OPG (1 Jul 1936) Stark BDC.
33. Stark to the OPG (7 Nov 1937) Stark BDC.
34. Gürtner to Hess (30 Jan 1936) Stark BDC.
35. Kleinert in Olff-Nathan (1993), 160.
36. See Kershaw, (1987), 83–104.
37. See Beyerchen, (1977), 15–78.
38. Fischer, (1991), 540–41.
39. Albrecht and Hermann, (1990); also see Macrakis, (1993).
40. Also see chapters 4 and 6.
41. Also see chapter 8.
42. Stark to Lenard (3 Feb 1933), in Kleinert, (1980), 35.
43. Lenard to Hitler (21 Mar 1933), in Beyerchen, (1977), 100.
44. Laue to Gerlach (25 Jul 1933), in Heinrich and Bachmann (1989), 73.
45. Stark to Frick (4 Nov 1933), in Kleinert, (1980), 43.
46. Stark to Lenard (3 May 1933), in Kleinert, (1980), 36.
47. Lenard to Stark (8 May 1933), in Kleinert, (1980), 36.
48. Stark to Lenard (27 Aug 1933), in Kleinert, (1980), 36.
49. Laue, (1947).

50. Laue, (1961).
51. Heilbron, (1986), 160.
52. Beyerchen, (1977), 114–15.
53. Hoffmann in Albrecht (1993), 122–24; Stark to Siebert (28 May 1934) 15.19 67/39 *Bundesarchiv Potsdam* (henceforth BAP).
54. Hoffmann in Albrecht (1993), 121–31; Stark to Siebert (28 May 1934) 15.19 67/39 BAP.
55. Stark to Ministry of the Interior (henceforth RIM) (26 Oct 1933) 15.01 26770/110 BAP.
56. RIM, "Aktenvermerk," (17 Nov 1933) 15.01 26770/179–81 BAP.
57. Donnevert (RIM), "Aktenvermerk," (Nov 1933) 15.01 26770/174 BAP.
58. Noakes and Pridham, (1990), 203–11; Kershaw, (1991), 1–15; Kershaw, (1993), 48–58; Renneberg and Walker, (1993), 1–11.
59. Stark to RIM (5 Dec 1933) 15.01 26770/219 BAP.
60. Hoffmann in Albrecht (1993), 126.
61. Stark to REM (28 Nov 1933) 15.01 26774/252 BAP.
62. Stark, (24 Feb 1934).
63. Stark, (21 Apr 1934).
64. Stark to RIM (24 Apr 1934) 15.19 65/328 BAP.
65. *Nationalsozialistes Lehrerbund* (henceforth NSLB) to Stark (1 Nov 1934) 15.19 65/169 BAP.
66. Stark to NSLB (8 Nov 1934) 15.19 65/171 BAP.
67. Stark to Hertz (13 Feb. 1936) 15.19 73/195 BAP.
68. Stark to Goebbels (23 Aug 1934) 15.19 65/276 BAP.
69. Beyerchen, (1977), 118.
70. Reprinted in Stark, (15 Jul 1937).
71. Stark to Goebbels (23 Aug 1934) 15.19 65/276 BAP.
72. Beyerchen, (1977), 117.
73. Stark to Lammers (5 Mar 1935) 15.19 68/80 BAP.
74. Stark to Schumann (3 Jul 1935) 15.19 67/132 BAP.
75. Beyerchen, (1977), 117.
76. Stark to Lenard (23 Jun 1934), in Kleinert (1980), 35.
77. Richter, (1973), 204.
78. Hermann, (1979), 83.
79. RIM to REM (Aug 1934) 15.01 26770/227 BAP.
80. See Beyerchen, (1977), 117–22.
81. Planck to Stark (31 Jan 1935) 15.19 65/359–61 BAP.
82. See Cassidy, (1991), 346–414.

83. Gall to Stark (3 Jul 1934) 15.19 65/255 BAP.
84. Stark to Gall (7 Jul 1934) 15.19 65/ 256 BAP.
85. Lenard to Stark (5 Dec 1934), in Hoffmann, (1989), 191.
86. Stark to Lenard (20 Dec 1934), in Hoffmann, (1989), 7.
87. Vahlen (REM) to Stark (29 Aug 1934) 15.19 67/188 BAP.
88. Lenard, (1935).
89. Richter, (1980), 117.
90. Kleinert (1979), 508.
91. Hoffmann, (1989), 192.
92. Beyerchen, (1977), 143.
93. Menzel, (29 Jan 1936).
94. Stark, (28 Feb 1936).
95. "Liste der Teilnehmer am Reichslager der Fachabteilung Physik, Darmstadt, von 14.-17.2.36," 15.19 68/131–42 BAP.
96. Willi Menzel, "Bericht über das Lager der Reichsfachabteilung Physik in Darmstadt vom 14. bis 17. Februar des Jahres" 15.19 68/131–42 BAP.
97. Stark to Wagner (24 Jul 1934) Stark BDC.
98. Wagner to Gürtner (26 Apr 1935) Stark BDC.
99. Gürtner to Hess (30 Jan 1936) Stark BDC.
100. Nippold to Bouhler (17 Feb 1936) Stark BDC.
101. Wagner to Gaugericht München-Oberbayern (22 Feb 1936) Stark BDC.
102. Nippold to Görlitzer (4 Mar 1936) Stark BDC.
103. Wagner to Gürtner (11 Mar 1936) Stark BDC.
104. Stark to OPG (29 Jun 1936).
105. Stark to OPG (1 Jul 1936) Stark BDC.
106. Gaugericht Gross-Berlin to OPG (18 Mar 1936) Stark BDC.
107. OPG to Gaugericht Gross-Berlin (16 May 1936) Stark BDC; Görlitzer to Wagner (7 Jul 1936) Stark BDC.
108. Der Stellvertreter des Führers to OPG (26 Nov 1937) Stark BDC.
109. Rust to Hitler (3 Feb 1936), in Albrecht and Hermann (1990), 382.
110. Stark to Lenard (3 Feb 1936) in Kleinert (1980), 36.
111. Stark to Lenard (11 Apr 1936) in Kleinert (1980), 37.
112. Lenard to Stark (24 Apr 1936) in Kleinert (1980), 38.
113. Stark to Lenard (29 Apr 1936) in Kleinert (1980), 38.
114. Beyerchen, (1977), 121.
115. Reischle to Himmler (24 Sep 1936) Stark BDC.
116. Stark to Mentzel (13 Feb 1937) 15.19 70/149–50 BAP.

117. Beyerchen, (1977), 121–22.
118. Stark to Lenard (12 Nov 1936) in Kleinert (1980), 38.
119. Noakes and Pridham, (1990), 167–87.
120. Wirz to Stark (18 Nov 1936) 15.19 71/326 BAP.
121. Ziegler to Stark (21 Nov 1936) 15.01 71/350 BAP.
122. Bühler to Stark (26 Nov 1936) 15.19 69/52 BAP.
123. Mentzel to Stark (4 Jan 1937) 15.19 70/144 BAP.
124. Stark to Mentzel (8 Jan 1937) 15.19 70/148 BAP.
125. Mentzel to Stark (18 Feb 1937) 15.19 70/153 BAP.
126. OPG to Stark (25 Jan 1937) Stark BDC.
127. OPG, "Vernehmungsprotokoll," (1 Feb 1937) Stark BDC.
128. Stark to OPG (4 Feb 1937) Stark BDC.
129. Volkmann (OPG), "Aktennotiz," (13 Mar 1937) Stark BDC.
130. Wagner to OPG (31 Mar 1937) Stark BDC.
131. Stark to Bewilogua (24 Jun 1937) 15.19 69/74 BAP.
132. Debye to Stark (6 Dec 1937) 15.19 69/84 BAP.
133. Bayer. Staatsmin. für Unterricht und Kultur to REM (9 Mar 1937) Heisenberg BDC.
134. Wesch to Stark (19 Jun 1937) 15.19 71/301 BAP.
135. REM to Stark (22 Jun 1937) 15.19 71/53 BAP.
136. Heisenberg to Spectabilität, Uni. Leipzig (17 Jul 1937) Uk. H185 IV, 45, Archives of the Humboldt University, Berlin (henceforth HUB).
137. "Weisse Juden" (15 Jul 1937).
138. "Weisse Juden" (15 Jul 1937).
139. Richter, (1980).
140. Hund to Rektor der Uni. Leipzig (henceforth RUL) (16 Jul 1937) Uk. H185 IV, 44 HUB.
141. Stark, (15 Jul 1937).
142. *Nature* to Stark (13 Oct 1937) 15.19 70/189 BAP.
143. Stark to *Nature* (15 Oct 1937) 15.19 70/190 BAP.
144. *Nature* to Stark (20 Oct 1937) 15.19 70/191 BAP.
145. Beyerchen, (1977), 77.
146. Rügemer, (1937).
147. Stark to Ziegler (4 Jan 1938) 15.19 71/363 BAP.
148. Stark, (30 Apr 1938).
149. See Kleinert (1979).
150. Kleinert in Olff-Nathan (1993), 162–63.
151. Himmler to Heisenberg (21 Jul 1938) Goudsmit, American Institute of Physics (henceforth AIP).

152. Himmler to Heydrich (21 Jul 1938) Goudsmit AIP.

153. OPG to Stark (14 Oct 1937) Stark BDC.

154. Oberlandesgerichts München to OPG (26 Jan 1937) Stark BDC; Gaugericht München-Oberbayern to OPG (21 Jun 1937) Stark BDC; Gaugericht München-Oberbayern to OPG (1 Nov 1937) Stark BDC.

155. Stark to OPG (26 Oct 1937) Stark BDC.

156. Stark to OPG (7 Nov 1937) Stark BDC.

157. Stark to OPG (23 Dec 1937) Stark BDC.

158. Schneider to Hess (16 Nov 1937) Stark BDC.

159. Hansen to OPG (11 Jan 1938) Stark BDC.

160. "In Sachen des Prof. Dr. Johannes Stark," (17 Jan 1938) Stark BDC; for Stark's fight with Wagner, see also Stark, (1987), 118–22.

161. For the "Sommerfeld Succession," see Cassidy (1991), 346–414.

162. Stark to Thüring (27 Jan 1938) 15.19 71/206 BAP.

163. Thüring to Stark (1 Feb 1938) 15.19 71/207 BAP.

164. Müller to Stark (20 Apr 1934) 15.19 66/69 BAP.

165. Stark to Langer (23 Jul 1937) 15.19 70/92 BAP.

166. Müller, (1939).

167. Beyerchen, (1977), 166.

168. Beyerchen, (1977), 171.

169. Kleinert in Olff-Nathan, (1993), 151.

170. Hentschel and Renneberg, (1994).

171. Beyerchen, (1977), 177–79, 192.

172. Heisenberg to Sommerfeld (4 Dec 1940) Sommerfeld, German Museum, Munich (henceforth DMM).

173. Gaupersonalamt to Gauschulungsamt im Hause (28 Oct 1942) Tomaschek BDC.

174. Heisenberg to Sommerfeld (4 Dec 1940) Sommerfeld DMM.

175. For Einstein attacked as a plagiarist, see Hentschel, (1990), 150–61, and Heisenberg, (1943).

176. See chapters 6 and 7.

177. Hönl to Heisenberg (17 Nov 1942) Heisenberg, in Max Planck Institute for Physics, Munich (henceforth MPI).

178. Sommerfeld to the Rektorat der Uni. München (1 Sep 1940) Sommerfeld DMM.

179. Prandtl, "Über die theoretische Physik," (28 Apr 1941) Ludwig Prandtl Papers, Göttingen (henceforth LPG).

180. See Beyerchen (1977) and Walker, (1989).

181. REM to Müller (21 Aug 1941) Sommerfeld DMM.

182. Stark to Müller (11 Apr 1941) Sommerfeld DMM.

183. Kleinert (1979), 509.

184. Müller to Rektor Wüst (23 Jun 1941) Sommerfeld DMM.

185. Müller to Kelter (11 Dec 1941) Sommerfeld DMM.

186. Müller to the Rektor (28 Jun 1942) Sommerfeld DMM.

187. Müller to Bergdolt (30 Sep 1942) Sommerfeld DMM.

188. Müller to Stark (24 Apr 1943) Sommerfeld DMM.

189. Selmayer to the Headquarters of General Eisenhower (2 May 1945) Sommerfeld DMM.

190. Müller to Stark (15 Apr 1944) Sommerfeld DMM.

191. Glaser to Stark (1 Feb 1938) 15.19 69/203 BAP.

192. Dingler to unknown addressee (3 Apr 1938) Glaser BDC.

193. Information in the BDC; Glaser joined in February of 1932.

194. Kanzlei des Führers to the Reichsleitung der NSDAP (5 Apr 1944) Glaser BDC.

195. Glaser, "Geist," (1939), 175.

196. *Beiseitigung*.

197. Glaser, "Feinstruktur," (1939), 329.

198. Thüring to Müller (24 Jun 1941) Sommerfeld DMM.

199. Finkelnburg to Gerlach (28 Jan 1941) in Heinrich and Bachmann, (1989) 80.

200. SA Gruppe Hochland to Kanzlei des Führers (5 Dec 1942) Glaser BDC; Glaser to Müller (29 Sep 1941) Sommerfeld DMM; Glaser to Müller (18 Nov 1941) Sommerfeld DMM.

201. Müller to Geisler (18 Sep 1942) Sommerfeld DMM.

202. Verwaltungsausschuss der Uni. München to Müller (29 Sep 1942) Sommerfeld DMM.

203. Glaser to Müller (10 Oct 1942) Sommerfeld DMM.

204. Kleinert in Olff-Nathan, (1993), 161–2; see Nagel, (1991).

205. Renneberg and Walker in Renneberg and Walker, (1993), 1–11.

206. Stark to Lenard (20 Apr 1934), in Kleinert (1980), 35.

207. Kleinert in Olff-Nathan, (1993), 162.

208. Kreisleiter Freisling, "Bestätigung," (9 Sep 1936) Stark BDC; SA Sturmbannführer, "Bescheinigung über die Tätigkeit in der S.A. und Partei," (14 Sep 1936) Stark BDC.

209. Stark, (1987), 135–37.

210. Kleinert, (1983), 14.

211. Kleinert, (1983), 16–18.

212. Beyerchen, (1977), 198, 206–7.

213. Kleinert (1983), 22; Berufungskammer München to Heisenberg (18 May 1949) Heisenberg MPI.
214. Heisenberg to Berufungskammer München (24 May 1949) Heisenberg MPI; Kleinert, (1983), 22.
215. Kleinert, (1983), 22.
216. Heisenberg to Berufungskammer München (24 May 1949) Heisenberg MPI; Kleinert, (1983), 22.
217. Stark, (1987), IV; Stark, (1947); see also Stark, (1987), 123–31.
218. Beyerchen, (1977), 203.
219. Beyerchen, (1977), 1.
220. Beyerchen, (1977), 201.
221. I do not mean to imply that Beyerchen's work is apologia, or that he presents a simplistic portrayal of science under Hitler. Although his book is mainly devoted to Lenard, Stark, and *Deutsche Physik*, he also provides an insightful and critical analysis of the physics community in general.
222. Dieter Hoffmann to the author (15 Jan, 11 Mar 1993).
223. Haberer, (1969), 165.
224. Richter (1980), 130.
225. Kleinert (1979), 513.
226. Grau, Schlicker, and Zeil, (1979), 9, 11.
227. Grau, Schlicker, and Zeil, (1979), 172–5.
228. See for example Haberer, (1969), 103–84, here 144.
229. Heilbron, (1986), 61–62.
230. Heilbron, (1986), 77–78.
231. For Planck's relationship with National Socialism, see Heilbron, (1986) and Albrecht in Albrecht (1993).
232. Heilbron, (1986), 155; for the Einstein affair in general see Grau, Schlicker, and Zeil, (1979), 7–12.
233. Letter from Planck to Einstein on March 19, quoted in Heilbron, (1986), 155.
234. Letter from von Laue to Einstein, quoted in Mehrtens, (1993), 327.
235. Heilbron, (1986), 158; Einstein's letter was written on April 12.
236. "Sitzung der Gesamt-Akademie" (henceforth SGA) (30 Mar 1933) II–V, 102, Archives of the Berlin–Brandenburg Academy of Sciences, Berlin (henceforth AdW).
237. Heilbron, (1986), 157; Heymann's statement was released on April 1.
238. Noakes and Pridham, (1990), 523–25; also see chapters 2 and 8.
239. Grau, Schlicker, and Zeil, (1979), 9.

240. SGA (6 Apr 1933) II–V, 102 AdW.
241. Heilbron, (1986), 159.
242. Grau, Schlicker, and Zeil, (1979), 11.
243. SGA (6 Apr 1933) II–V, 102 AdW.
244. Haberer, (1969), 116.
245. Grau, Schlicker, and Zeil, (1979), 12.
246. Eckert, (1992), 40, 44–8.
247. *Extraordinarius.*
248. Heilbron, (1986), 115, 120.
249. For the Stark Affair in general, see Grau, Schlicker, and Zeil, (1979), 172–75.
250. Grau, Schlicker, and Zeil, (1979), 173; Hoffmann, (1982), 97–8.
251. Beyerchen, (1977), 115–16; Grau, Schlicker, and Zeil, (1979), 172–75; Heilbron, (1986), 160–61.
252. Laue, (1947).
253. Hoffmann, (1989).
254. "Sitzung der phys.-math. Klasse" (henceforth SPM) (25 Feb 1937) II–V, 138 107 AdW.
255. Vahlen to Heisenberg (8 Sep 1942) Heisenberg MPI; Heisenberg to Vahlen (10 Sep 1942) Heisenberg MPI.
256. Beyerchen, (1977), 64–66.
257. "Verzeichnis der Dozenten für Mathematik an der Universität Berlin und deren Verurteilung in politischer und wissenschaftlicher Hinsicht," Bieberbach BDC.
258. Laue to Stark (2 Dec 1936) 15.19 70/95 BAP.
259. Also see Grau, Schlicker, and Zeil, (1979).
260. Noakes and Pridham, (1990), 220–32, esp. 223–25.
261. Noakes and Pridham, (1990), 523–25, 553–65.
262. Pfundtner (RIM) to die Landesregierungen (11 Jun 1934) II–IV, 9 AdW; REM to die nachgeordneten Reichsdienststellen (henceforth NRD) (14 Aug 1934) II–IV, 9 AdW.
263. RIM to die Landesregierungen und die Herrn Reichsstatthalter (21 Jun 1934) II–IV, 9 AdW; REM to die NRD (6 Jul 1934) II–IV, 9 AdW.
264. Der preussische Justizminister, "Strafsachen" (6 Jul 1934) II–IV, 9 AdW.
265. REM to NRD (23 Mar 1935) II–IV, 9 AdW.
266. PAW to its members (20 Sep 1935) II–IV, 9 AdW.
267. SGA (24 Oct 1935) II–V, 103 AdW.
268. SGA (5 Dec 1935) II–V, 103 AdW.

269. REM to NRD (30 Aug 1937) II–IV, 10 AdW.

270. RIM to die Obersten Reishsbehörden (18 Dec 1937) II–IV, 10 AdW.

271. "Auszug aus dem Protokoll der Sekretariatssitzung" (13 Jul 1933) II–IV, 21/43 AdW.

272. See below.

273. "Verhandlungsniederschrift der Vertretersammlung des Verbandes der deutschen Akademien am Freitag, den 26. Mai 1933 in Leipzig" (26 May 1933) II–XII, 9 AdW.

274. Heidelberg Academy to PAW (22 Dec 1935) II-XII, 10 AdW.

275. "Auszug aus dem Protokoll der Sekretariatssitzung" (26 Sep 1935) II–IV, 9 AdW.

276. REM to NRD (30 Jan 1936) II–IV, 10 AdW.

277. REM to NRD (11 Jan 1936) II–IV, 10 AdW; PAW to REM (20 Jan 1936) II–IV, 10 AdW.

278. PAW to REM (2 Jan 1936) II–IV, 10 AdW; REM to NRD (9 Apr 1937) II–IV, 10 AdW.

279. SGA (8 Feb 1934) II–V, 103 AdW.

280. SGA (26 Sep 1935) II–V, 103 AdW.

281. See chapter 2.

282. Frick & Goebbels, "Trauererlass zum Ableben des Reichspräsidenten von Hindenburg von 2. August 1934" (2 Aug 1934) II–IV, 9 AdW.

283. Paul Harteck, "Diensteid," (19 Dec 1934) HSWD- & PH I206 Bd. 1 HSA.

284. Frick (RIM) to die Obersten Reichsbehörden (21 Aug 1934) II–IV, 9 AdW.

285. Rust (REM) to NRD (12 Jul 1935) II–IV, 9 AdW.

286. See chapter 2.

287. REM, "Betrifft: Ernennungen und Beförderungen von Beamten, Angestellten, und Arbeitern" (6 Nov 1934) II–IV, 9 AdW.

288. SPM (30 Jan 1936) II–V, 138/57 AdW.

289. PAW to Beamten, Angestellten und Mitarbeiter der Akademie (26 Sep 1936) II–IV, 10 AdW.

290. SGA (10 Jan 1935) II–V, 103 AdW.

291. SGA (30 Apr 1936) II–V, 103 AdW.

292. SGA (14 May 1936) II–V, 103 AdW.

293. SPM (21 Feb 1935) II–V, 138/9 AdW.

294. Grau, Schlicker, and Zeil, (1979), 48.

295. This information comes from Bieberbach, BDC.

296. Mehrtens, in Renneberg and Walker, (1993), 291–311 & 406–8, here 298–99; also see Mehrtens, in Renneberg and Walker, (1993), 324–38 & 411–13.

297. SGA (24 Oct 1935) II–V, 103 AdW.

298. For Ludendorff and National Socialism, see Hentschel, *Turm*, (1992).

299. SPM (13 Feb 1936) II–V, 138/61 AdW.

300. SPM (27 Feb 1936) II–V, 138/66 AdW.

301. SPM (6 Jun 1935) II–V, 138/27 AdW.

302. SGA (15 Oct 1936) II–V, 103 AdW.

303. SGA (30 Apr 1936) II–V, 103 AdW.

304. SGA (14 May 1936) II–V, 103 AdW.

305. SGA (6 Feb 1936) II–V, 103 AdW.

306. SGA (20 Feb 1936) II–V, 103 AdW.

307. SGA (5 Mar 1936) II–V, 103 AdW.

308. SGA (4 Feb 1937) II–V, 104 AdW.

309. SGA (18 Feb 1937) II–V, 104 AdW.

310. "Aktenvermerk" (25 Feb 1937) II–I, 12/11 AdW; also see Grau, Schlicker, and Zeil, (1979), 62.

311. REM to PAW (17 Apr 1937) II–I, 12/17 AdW.

312. Wacker (REM) to das Sächisiche Ministerium für Volksbildung (10 Feb 1937) II-XII, 10 AdW; REM to PAW (10 Feb 1937) II–I, 12/20 AdW.

313. "Aktenvermerk" (25 Feb 1937) II–I, 12/11 AdW.

314. PAW to REM (1 Mar 1936) II–I, 12/24 AdW.

315. "Protokoll der Ausserordentlichen Kartellsitzung der Deutschen Akademien, Berlin am 24. April 1937," (24 Apr 1937) II-XII, 10 AdW.

316. SGA (29 Apr 1937) II–V, 104 AdW.

317. SGA (27 May 1937) II–V, 104 AdW.

318. von Ficker (PAW) to Kees (8 May 1937) II-XII, 10 AdW.

319. SPM (25 Feb 1937) II–V, 138/107 AdW.

320. SPM (11 Feb 1937) II–V, 138/104 AdW.

321. Information in Vahlen, BDC; for Vahlen also see Siegmund-Schultze, (1984).

322. Siegmund-Schultze, (1984), 19–21.

323. Siegmund-Schultze, (1984), 27.

324. Vahlen to Heisenberg (8 Sep 1942) Heisenberg MPI; Heisenberg to Vahlen (10 Sep 1942) Heisenberg MPI.

325. Siegmund-Schultze, (1984).

326. Lenard to Stark (3 Apr 1936) in Kleinert, (1980), 36.

327. Lenard to Vahlen (6 Apr 1936) in Kleinert, (1980), 37.

328. Stark to Lenard (6 Apr 1936) in Kleinert, (1980), 37.
329. Stark to Lenard (11 Apr 1936) in Kleinert, (1980), 37.
330. Siegmund-Schultze, (1984), 22; the two mathematicians were G. Doetsch and Karl Willi Wagner.
331. SGA (15 Apr 1937) II–V, 104 AdW.
332. Bieberbach to die physikalische-mathematische Klasse der PAW (19 Apr 1937) II–I, 12/1 AdW.
333. SPM (13 May 1937) II–V, 139/6 AdW.
334. SGA (10 Jun 1937) II–V, 104 AdW.
335. SGA (24 Jun 1937) II–V, 104 AdW.
336. REM to Vahlen (25 Oct 1937) Vahlen BDC.
337. SS Sicherheitsdienst to die Personalkanzlei des Reichsführers-SS (29 Jun 1936) Vahlen BDC.
338. Rust (REM) to PAW (8 Oct 1938) II–I, 13/16 AdW.
339. Noakes and Pridham, (1990), 536–37.
340. Rust (REM) to PAW (8 Oct 1938) II–I, 13/16 AdW.
341. SGA (13 Oct 1938) II–V, 104 AdW.
342. PAW to REM (14 Oct 1938) II–I, 13/16 AdW.
343. Spielvogel, (1992), 273–4; Noakes and Pridham, (1990), 553–565.
344. REM to PAW (22 Nov 1938) II–I, 13/27 AdW.
345. Meissner, Vahlen, Bieberbach, Kraft, and Grapow to PAW (before 1 Dec 1938) II–I, 13/33 AdW.
346. SGA (1 Dec 1938) II–I, 13/34 AdW.
347. SGA (1 Dec 1938) II–V, 104 AdW; SGA (1 Dec 1938) II–I, 13/36 AdW.
348. SGA (15 Dec 1938) II–V, 104 AdW.
349. PAW to REM (22 Dec 1938) II–I, 13/50 AdW.
350. REM to PAW (24 Dec 1938) II–I, 13/59 AdW.
351. SGA (12 Jan 1939) II–V, 104 AdW.
352. Noakes and Pridham, (1990), 697–99.
353. SPM (19 Jan 1939) II–V, 139/64–66 AdW.
354. REM to PAW (17 Sep 1938) II–III, 107/1 AdW.
355. SPM (19 Jan 1939) II–V, 139/64–66 AdW.
356. SGA (26 Jan 1939) II–V, 104 AdW.
357. SGA (26 Jan 1939) II–V, 104 AdW.
358. SGA (2 Mar 1939) II–V, 104 AdW.
359. SGA (22 Mar 1939) II–V, 104 AdW.
360. SGA (15 Jun 1939) II–V, 104 AdW.
361. SGA (15 Jun 1939) II–V, 104 AdW.
362. SGA (29 Jun 1939) II–V, 104 AdW.

363. Heilbron, (1986), 173.

364. For the international context during the Weimar Republic and early years of the Third Reich, see Schroeder-Gudehus, (1972); Schroeder-Gudehus, (1973); and especially Schroeder-Gudehus, (1978).

365. SGA (24 Jun 1937) II–V, 104 AdW.

366. SGA (23 Jun 1938) II–V, 104 AdW.

367. SGA (20 Oct 1938) II–V, 104 AdW.

368. SGA (17 Mar 1938) II–V, 104 AdW.

369. SGA (2 Nov 1939) II–V, 104 AdW.

370. SPM (18 Apr 1940) II–V, 139/110 AdW.

371. Reichstauschstelle (REM) to PAW (4 Dec 1936) II-XVI, 81/1 AdW; Jürgens to PAW (31 Jan 1938) II-XVI, 81/63 AdW.

372. SPM (7 Dec 1939) II–V, 139/100 AdW.

373. PAW to REM (9 Jan 1940) II-XVI, 81 AdW.

374. "Sitzung der math.-naturw. Klasse" (7 Nov 1940) II–V, 139/121 AdW.

375. Bieberbach to REM (30 Sep 1941) II-XVI, 58 AdW; REM to Vahlen (19 Jan 1942) II-XVI, 58 AdW.

376. SGA (25 Jul 1940) II–V, 104 AdW.

377. SGA (2 Nov 1939) II–V, 104, AdW.

378. SGA (16 Sep 1940) II–V, 104 AdW.

379. Staats- und Universitäts-Bibliothek Posen to PAW (27 Nov 1940) II–XVI, 73 AdW; PAW to Staats- und Universitäts-Bibliothek Posen (2 Dec 1940) II–XVI, 73 AdW.

380. REM to PAW (6 Dec 1940) II–XVI, 73 AdW.

381. Staats- und Universitäts-Bibliothek Posen to PAW (2 Jan 1941) II–XVI, 73 AdW.

382. Staatsbibliothek (Amt des Generalgourverneurs) to PAW (17 Feb 1941) II–XVI, 73 AdW.

383. SGA (24 Apr 1941) II–V, 104 AdW.

384. Kungl. Vitterhets Histoire och Antikvitets Akademien to PAW (7 Dec 1943) II–XIV 87/116 AdW.

385. SGA (13 Jan 1944) II–V, 104 AdW.

386. SGA (24 Feb 1944) II–V, 104 AdW.

387. Rörig to Schück (7 May 1944) II–XIV, 87/146 AdW.

388. SPM (8 Jun 1939) II–V, 139/86 AdW.

389. SGA (22 Jun 1939) II–V, 104 AdW; SGA (29 Jun 1939) II–V, 104 AdW.

390. SGA (25 Mar 1943) II–V, 104 AdW.

391. "Sitzung der math.-naturw. Klasse" (5 Mar 1941) II–V, 139/162 AdW.

392. SGA (29 Feb 1940) II–V, 104 AdW.

393. SGA (13 Jul 1939) II–V, 104 AdW.

394. SGA (27 Nov 1941) II–V, 104 AdW.

395. SGA (19 Nov 1942) II–V, 104 AdW.

396. REM to Vahlen (16 Oct 1939) Vahlen BDC; Vahlen to Personalamt der Schutzstaffeln (Berlin) (9 Nov 1939) Vahlen BDC; Vahlen to Personalamt der Schutzstaffeln (Berlin) (1 Mar 1940) Vahlen BDC; Personalamt der Schutzstaffeln (Berlin) to Vahlen (5 Mar 1940) Vahlen BDC.

397. Vahlen to Himmler (19 Feb 1943) Vahlen BDC; "Vermerk" (23 Feb 1943) Vahlen BDC; Brandt to Sievers (25 Feb 1943) Vahlen BDC.

398. Sievers (Ahnenerbe) to Brandt (13 Mar 1943) Vahlen BDC.

399. Himmler to Vahlen (25 Mar 1943) Vahlen BDC.

400. Brandt to von Herrff (10 Oct 1943) Vahlen BDC.

401. SS-Personalhauptamt to Vahlen (3 Feb 1944) Vahlen BDC.

402. Brandt to Vahlen (14 Feb 1944) Vahlen BDC.

403. SGA (24 Jul 1941) II–V, 104 AdW.

404. SGA (25 Nov 1943) II–V, 104 AdW.

405. SGA (16 Dec 1943) II–V, 104 AdW.

406. "Sitzung der math.-naturw. Klasse" (6 Jan 1944) II–V, 139/212 AdW.

407. SGA (9 Mar 1944) II–V, 104 AdW.

408. Fakultät für Naturwissenschaften to TU Wien (3 Apr 1944) Vahlen BDC; TU Wien to REM (8 May 1944) Vahlen BDC.

409. SGA (6 Jul 1944) II–V, 104 AdW.

410. SGA (1 Feb 1945) II–V, 104 AdW.

411. SGA (6 Jun 1945) II–V, 104 AdW.

412. SGA (21 Jun 1945) II–V, 104 AdW.

413. SGA (14 Jun 1945) II–V, 104 AdW.

414. SGA (28 Jun 1945) II–V, 104 AdW.

415. "Aufzeichnung" (~Jun 1945) II–I, 14/16 AdW.

416. SGA (12 Jul 1945) II–V, 104 AdW.

417. SGA (6 Dec 1945) II–V, 104 AdW.

418. SGA (2 Aug 1945) II–I, 14/18 AdW.

419. SGA (23 Aug 1945) II–V, 104 AdW.

420. SGA (13 Dec 1945) II–V, 104 AdW.

421. SGA (13 Dec 1945) II–V, 104 AdW.

422. SGA (20 Dec 1945) II–V, 104 AdW; Stroux to Magistrat der Stadt Berlin Abteilung Wissenschaft (27 Dec 1945) II–I, 14/31 AdW.

423. Deutsche Akademie der Wissenschaften to various institutes (19 Apr 1945) II–IV, 10 AdW.

424. Deutsche Zentralverwaltung für Volksbildung to Akademie der Wissenschaften (4 Oct 1946) II–IV, 10 AdW.

425. Siegmund-Schultze, (1984), 31, footnote 19.

426. Haberer, (1969), 145.

427. See chapter 10.

428. Das Preussische Min. für Wissenschaft, Kunst and Volksbildung to Rektor der Uni. Berlin (henceforth RUB) (23 May 1933) PF 162, 62 HUB.

429. RUB to die Mitglieder des Lehrkörpers (26 Jan 1934) PF 162, 66 HUB; SGA II–V, 103 AdW.

430. German Foreign Office (henceforth AA) to sämtliche Reichsministerien (30 Sep 1934) PF 162, 82 HUB.

431. REM to Rektoren aller deutschen Hochschulen (9 Jan 1935) PF 1408, 45 HUB.

432. REM to Rektoren aller deutschen Hochschulen (26 Jan 1937) PF 1408, 41 HUB.

433. SGA II–V, 103 AdW.

434. SGA II–V, 103 AdW.

435. Sächsisches Min. für Volksbildung (henceforth SMV) to die phil. Fak. der Uni. Leipzig, (5 Jan 1929) Uk. H185 III, 20 HUB; SMV to die phil. Fak. der Uni. Leipzig (2 Feb 1929) Uk. H185 III, 24 HUB.

436. SMV to Heisenberg (21 May 1932) Uk. H185 III, 35 HUB.

437. Dekane der Uni. Leipzig to Heisenberg (10 Nov 1933) Uk. H185 III, 36 HUB; Heisenberg to phil. Fak. (29 Nov 1933) Uk. H185 III, 37 HUB.

438. Heisenberg to Dekan der phil. Fakultät (Köbe) (29 May 1934) Uk. H185 III, 38 HUB; Dekan to Heisenberg (1 Jun 1934) Uk. H185 III, 39 HUB.

439. Heisenberg to RUL (19 Feb 1936) Uk. H185 III, 41 HUB; REM to RUL (30 Mar 1936) Uk. H185 III, 42 HUB.

440. Heisenberg to RUL (30 May 1936) Uk. H185 III, 44 HUB.

441. Hoffmann, (1988); for Jordan also see Wise, (1993) and Beyler, (1994).

442. Heisenberg to RUL (3 Apr 1937) REM 2944, 15 BAP.

443. RUL to REM (22 Apr 1937) REM 2944, 14 BAP; Dozentenschaft der Uni. Leipzig to RUL (13 Apr 1937) REM 2944, 16 BAP.

444. Dozentenschaft der Uni. Leipzig to RUL (13 Apr 1937) REM 2944, 16 BAP.

445. *Völkischer Beobachter* (29 Jan 1936); *Völkischer Beobachter* (29 Feb 1935); see chapters 2 and 3.

446. Heisenberg to RUL (16 Jul 1937) REM 2944, 112 BAP; RUL to REM (28 Jul 1937) REM 2944, 111 BAP; REM to RUL (12 Aug 1937) REM 2944, 113 BAP.

447. Rektor der Uni. Halle-Wittenberg to REM (15 May 1936) REM 2943, 433 BAP; Geiger to Rektor der Uni. Tübingen (23 May 1936) REM 2943, 442 BAP; Rektor der Uni. Tübingen, "Vermerk," (27 May 1936) REM 2943, 442 BAP; Döpel to Rektor der Uni. Würzburg (29 May 1936) REM 2943, 440 BAP; Rektor der Uni. Würzburg to Bayer. Staatsministerium für Unterricht und Kultur (3 Jun 1936) REM 2943, 441 BAP; REM to Rektor der Uni. Halle (9 Jun 1936) REM 2943, 434 BAP; REM to AA (23 Jun 1936) REM 2943, 448 BAP; REM to AA (27 Jun 1936) REM 2943, 457 BAP.

448. Geiger to Rektor der Uni. Tübingen (17 Jun 1936) REM 2943, 454 BAP.

449. Rektor der Uni. Kiel to REM (19 Jun 1936) REM 2943, 462 BAP; REM to Rektor der Uni. Kiel (6 Jul 1936) REM 2943, 471 BAP; Gauleitung Schleswig-Holstein to Rektor der Uni. Kiel (23 Jun 1936) REM 2943, 468 BAP; Rektor der Uni. Kiel to REM (25 Jun 1936) REM 2943, 469 BAP.

450. Sauter to Rektor der Uni. Göttingen (25 May 1936) REM 2943, 443 BAP; Sauter's membership in the NSDAP is documented in the BDC.

451. Rektor der Uni. Göttingen to REM (28 May 1936) REM 2943, 444 BAP; REM to Gestapo (Best) (23 Jun 1936) REM 2943, 451 BAP; REM, "Vermerk," 23 Jun 1936) REM 2943, 451 BAP.

452. REM to Universitätskurator Göttingen (29 Jun 1936) REM 2943, 464 BAP.

453. Regener to Rektor der TH. Stuttgart, (4 May 1936) REM 2943, 435 BAP.

454. REM to Württembergischer Kultusminister (11 Jun 1936) REM 2943, 437 BAP.

455. Württembergischer Kultusminister to REM (12 Jun 1936) REM 2943, 446 BAP.

456. REM to Württembergischer Kultusminister (23 Jun 1936) REM 2943, 448–49 BAP.

457. For both the period of relative calm during the Third Reich and "The Night of Broken Glass," see Noakes and Pridham, (1990), 547–65.

458. *Das Schwarze Korps* (15 Jul 1937); see chapters 2 and 3.

459. Heisenberg to Dekan der phil. Fak. der Uni. Leipzig (17 Jul 1937) Uk. H185 IV, 45 HUB.

460. Das Preussische Min. für Wissenschaft, Kunst and Volksbildung to PAW (28 Apr 1933) II–IV, 9 AdW.

461. REM to PAW (10 Sep 1936) II–IV, 10 AdW.
462. RUL to Reichsstatthalter in Sachsen (22 Jul 1937) Uk. H185 IV, 50 HUB.
463. Heisenberg to Dekan der Uni. Leipzig (22 Feb 1938) Uk. H185 IV, 42 HUB.
464. Heisenberg to Sommerfeld (14 Apr 1938) 1977–28/A, 136 (18 DMM).
465. See chapters 2 and 3.
466. Prandtl to Himmler (12 Jul 1938) LPG; for Prandtl also see Trischler, (1993).
467. Kreisleiter (Göttingen) to das Amt für Technik Gauamtsleitung (28 May 1937) Prandtl BDC.
468. Himmler to Prandtl (21 Jul 1938) Heisenberg BDC; Himmler to Heisenberg (21 Jul 1938) Goudsmit AIP; Himmler to Heydrich (21 Jul 1938) Goudsmit AIP.
469. Heisenberg to Himmler (23 Jul 1938) Goudsmit AIP; Heisenberg to Sommerfeld (23 Jul 1938) 1977–28/A, 136 (21 DMM).
470. Heisenberg to Sommerfeld (24 Nov 1938) 1977–28/A, 136 (?? DMM.
471. Prandtl to Heckmann (29 Nov 1938) Heckmann BDC.
472. SMV (Studentkowski) to RUL (Knick) (22 Dec 1938) Uk. H185 IV, 37 HUB.
473. Heisenberg to RUL (9 Jan 1939) Uk. H185 IV, 29 HUB.
474. RUL to Heisenberg (12 Jan 1939) Uk. H185 IV, 30 HUB; REM to RUL (21 Jan 1939) Uk. H185 IV, 31 HUB.
475. Werner Heisenberg, "Bericht über eine Vortragsreise nach Holland," (1 Feb 1939) Uk. H185 IV, 35 HUB.
476. Heisenberg to RUL (13 Apr 1939) Uk. H185 IV, 25 HUB.
477. RUL to REM (19 Apr 1939) H185 IV, 26 HUB.
478. REM to RUL (27 May 1939) Uk. H185 IV, 27 HUB; REM to Rektor der Montanistischen Hochschule in Leoben (11 Oct 1939) REM 2943, 410 BAP.
479. See chapters 2 and 3.
480. Erxleben to Parteikanzlei (9 Sep 1942) MA 116/5 Wissenschaft Heisenberg, Institut für Zeitgeschichte, Munich (henceforth IZG) and Borger to Parteikanzlei (9 Sep 1942) MA 116/5 Wissenschaft Heisenberg IZG; Heisenberg to Sommerfeld (30 Jan 1939) 1977–28/A, 136 (25 DMM; Heisenberg to Sommerfeld (15 Feb 1939) 1977–28/A, 136 (26 DMM; Heisenberg to Sommerfeld (3 Mar 1939) 1977–28/A, 136 (28 DMM; Heisenberg to Sommerfeld (13 May 1939) 1977–28/A, 136 (30 DMM).

481. SS to REM (26 May 1939) REM 2943, 370–71 BAP.
482. Also see Cassidy, (1991), 50–63.
483. SS to REM (26 May 1939) REM 2943, 370–71 BAP.
484. Prandtl to Himmler (12 Jul 1938) LPG; Himmler to Heisenberg (7 Jun 1939) Goudsmit AIP.
485. REM to PAW (13 Aug 1937) II–IV, 10 AdW.
486. Parteikanzlei to REM (12 Jan 1939) REM 2943, 373 BAP.
487. Parteikanzlei to REM (22 Jun 1939) REM 2943, 369 BAP.
488. Esau BDC.
489. REM to Esau (25 Aug 1939) Uk. H185 IV, 28 HUB.
490. Esau to REM (13 Jun 1939) REM 2943, 367 BAP.
491. REM to Parteikanzlei (25 Aug 1939) REM 2943, 391–92 BAP.
492. Heisenberg to Dekan der phil. Fak. der Uni. Leipzig (19 Sep 1939) Uk. H185 III, 45 HUB.
493. Heisenberg to Dekan der phil. Fak. der Uni. Leipzig (26 Sep 1939) Uk. H185 III, 46 HUB; Heisenberg to phil. Fak. der Uni. Leipzig (15 Jun 1940) Uk. H185 III, 47 HUB.
494. Heisenberg to Sommerfeld, (4 Sep 1939) 1977–28/A,136 (32 DMM).
495. Noakes and Pridham, (1990), 597–8; Spielvogel, (1992), 243.
496. REM, "Merkblatt," (1 Jun 1942) Heisenberg MPI; REM, "Merkblatt," (1 Mar 1943) Heisenberg MPI.
497. REM to RUB (3 Aug 1939) PF 1408, (149–50 HUB.
498. REM to RUB, (19 Jun 1941) PF 1483, 105 HUB.
499. REM, "Merkblatt," (1 Jun 1942) Heisenberg MPI.
500. REM to RUB (2 Jul 1941) PF 1483, 107 HUB; REM, "Merkblatt," (1 Jun 1942) Heisenberg MPI; REM to Rektor Uni. Berlin (6 Jul 1942) PF 1484, 33 HUB.
501. For the General Government, see Noakes and Pridham, (1990), 922–6.
502. REM to RUB (31 Oct 1939) PF 1483, 59 HUB.
503. REM to RUB (16 Jul 1942) PF 1484, 34 HUB.
504. REM to RUB (24 Sep 1942) PF 1484, 37 HUB.
505. REM to RUB (16 Jul 1942) PF 1484, 34 HUB.
506. Heisenberg to RUL (5 Nov 1940) Uk. H185 IV, 24 HUB.
507. "Memo," (8 Nov 1940) Uk. H185 IV, 24 HUB.
508. Heisenberg to RUL (4 Dec 1940) Uk. H185 IV, 23 HUB.
509. Dozentenführer to RUL (9 Dec 1940) Uk. H185 IV, 23 HUB.
510. RUL to REM (11 Dec 1940) Uk. H185 IV, 22 HUB.
511. REM to RUL (24 Jan 1941) Uk. H185 IV, 21 HUB.
512. REM to RUL (19 Mar 1941) Uk. H185 IV, 20 HUB.

513. Coblitz to Heisenberg (20 May 1941) Heisenberg MPI.
514. Mehrtens, (1986), 344.
515. Weinreich, (1946), 95–7.
516. REM to Frank (29 Jul 1941) REM 690, 67 BAP.
517. Heisenberg to Coblitz (6 Jun 1941) Heisenberg MPI.
518. RUL to REM (23 Jun 1941) Uk. H185 IV, 16 HUB.
519. RUL to REM (23 Jul 1941) Uk. H185 IV, 14 HUB.
520. REM to RUL (9 Sep 1941) Uk. H185 IV, 13 HUB.
521. REM to Frank (29 Jul 1941) REM 690, 67 BAP.
522. Der Bevollmächtigte des Deutschen Reiches, Denmark to AA (Berlin) (27 Mar 1941) REM 2943, 524–25 BAP.
523. AA, "Vermerk," (16 Apr 1941) REM 2943, 525 BAP.
524. Von Weizsäcker to the author (23 May, 13 Jun, 26 Jun, 1 Aug, and 5 Aug 1990).
525. REM to KWG (8 May 1941) REM 2943, 528 BAP; KWG to REM (20 May 1941) REM 2943, 529 BAP.
526. REM to AA (3 Jun 1941) REM 2943, 530 BAP.
527. Der Bevollmächtigte des Deutschen Reiches, Denmark to AA (27 Jun 1941) REM 2943, 537 BAP.
528. Von Weizsäcker to German Academic Exchange Service (henceforth DAAD) (22 Jun 1941) REM 2943, 538 BAP.
529. AA to REM (2 Aug 1941) REM 2943, 531 BAP.
530. Von Weizsäcker to Bohr (15 Aug 1941) Niels Bohrs General Correspondence, Niels Bohr Institute, Copenhagen (henceforth NBC).
531. REM to REM (WT) (18 Aug 1941) REM 2943, 539 BAP.
532. REM to AA (28 Aug 1941) REM 2943, 533 BAP [never sent].
533. REM, "Vermerk," (21 Aug 1941) REM 2943, 532 BAP.
534. REM, "Vermerk," (2 Sep 1941) REM 2943, 534–35 BAP.
535. For Mentzel's position, see SGA II–V, 104 AdW.
536. REM, "Vermerk," (2 Sep 1941) REM 2943, 534–35 BAP.
537. REM to Parteikanzlei, (around 2 Sep 1941) REM 2943, 535 BAP.
538. REM, "Vermerk," (11 Sep 1941) REM 2943, 536 BAP.
539. REM to RUL (9 Sep 1941) Uk. H185 IV, 13 HUB.
540. REM, "Vermerk," (11 Sep 1941) REM 2943, 536 BAP.
541. Werner Heisenberg, "Bericht," (23 Sep 1941) REM 2943, 547 BAP.
542. Carl Friedrich von Weizsäcker, "Bericht," (1 Oct 1941) REM 2943, 549–50 BAP.
543. Crowther, (1949), 108.
544. Feinberg, (1989).

545. Mott and Peierls, (1977), 236.

546. Der Bevollmächtigte des Deutschen Reiches, Denmark (Seelos) to AA (26 Sep 1941) REM 2943, 544–45 BAP.

547. AA to REM (27 Nov 1941) REM 2943, 557 BAP.

548. Burrin, (1994); see below.

549. Prandtl to Göring (28 Apr 1941) LPG; Ludwig Prandtl, "Über die theoretische Physik," (28 Apr 1941) LPG; Prandtl to Joos (29 May 1941) LPG; Joos to Prandtl (6 Jun 1941) LPG; Prandtl to Ramsauer (8 Jun 1941) LPG.

550. Prandtl to Ramsauer (31 Oct 1941) LPG.

551. Prandtl to Milch (13 Nov 1941) LPG.

552. Lucht to Prandtl (3 Dec 1941) LPG.

553. Ramsauer to Rust (20 Jan 1942) LPG.

554. Prandtl to Ramsauer (28 Jan 1942) LPG.

555. Finkelnburg to Heisenberg (6 May 1942) Heisenberg MPI; Heisenberg to Finkelnburg (22 May 1942) Heisenberg MPI; Heisenberg to Jordan (31 Jul 1942) Heisenberg MPI; Sommerfeld to Heisenberg (19 Oct 1942) Heisenberg MPI.

556. Schumann to Harteck, 4 Feb 42, Erich Bagge Papers, Kiel (henceforth EBK); President of the Reich Research Council to Harteck (21 Feb 42) EBK; Heisenberg to Rust (20 Feb 42) Heisenberg MPI.

557. Rust to Lorenz (12 Feb 42) 29–993 David Irving Microfilm Collection, AIP (henceforth Irving).

558. Goudsmit, (1983), 168–71.

559. Otto Hahn, "Die Spaltung des Urankerns," (26 Feb 1942) *G-150* German Reports, AIP.

560. Paul Harteck, "Die Gewinnung von schwerem Wasser," (26 Feb 1942) *G-154* German Reports, AIP.

561. Werner Heisenberg, "Die theoretischen Grundlagen für die Energiegewinnung aus der Uranspaltung," (26 Feb 1942), reprinted in Heisenberg, (1989), 517–21.

562. Otto Hahn, "Die Spaltung des Urankerns," (26 Feb 1942) *G-150* German Reports, AIP.

563. Paul Harteck, "Die Gewinnung von schwerem Wasser," (26 Feb 1942) *G-154* German Reports, AIP.

564. The figure is taken from a subsequent talk, Werner Heisenberg, "Die Energiegewinnung aus der Atomkernspaltung," (6 May 1943), reprinted in Heisenberg, (1989), 570–75.

565. Heisenberg, (1989), 517–21.

566. Heisenberg to Goudsmit, (5 Jan 48), Goudsmit AIP.

567. Heisenberg, (1989), 517–21.

568. Hahn's diary (26 Feb 42), 29–021 Irving, AIP.

569. Newspaper clipping (paper unknown) (27 Feb 42), "Physik und Landesverteidigung," Max Planck Society Archives, Berlin (henceforth MPGB).

570. Finkelnburg to Heisenberg (6 May 42) Heisenberg MPI.

571. Goebbels, (1948), 140.

572. Hans Meckel to the Editor, *Die Welt* (5 Jun 1991), 5.

573. Parteikanzlei to Härtle (8 Jul 1942) MA 116/5 Wissenschaft Heisenberg IZG; Erxleben to Parteikanzlei (9 Sep 1942) MA 116/5 Wissenschaft Heisenberg IZG; Borger to Parteikanzlei (9 Sep 1942) MA 116/5 Wissenschaft Heisenberg IZG.

574. Parteikanzlei to Härtle (8 Jul 1942) MA 116/5 Wissenschaft Heisenberg IZG; Forstmann (KWG) to Heisenberg (13 Oct 1942) Heisenberg MPI; Telschow to Mentzel (1 Mar 1943) Irving, AIP.

575. Heisenberg to Schmellenmeier (10 Jan 1944) Heisenberg MPI.

576. Borger to Parteikanzlei (9 Sep 1942) MA 116/5 Wissenschaft Heisenberg IZG.

577. Erxleben to Borger (10 Jul 1942) MA 116/5 Wissenschaft Heisenberg IZG; Erxleben to Parteikanzlei (9 Sep 1942) MA 116/5 Wissenschaft Heisenberg IZG.

578. Scherrer to Heisenberg (26 May 1942) Heisenberg MPI; Heisenberg to Dekan phil. Fak. der Uni. Leipzig (10 Jun 1942) Heisenberg MPI.

579. Stueckelberg to Heisenberg (20 Aug 1942) Heisenberg MPI; Studentenschaft der Universität Bern to Heisenberg (7 Sep 1942) Heisenberg MPI; Fischer to Heisenberg (23 Oct 1942) Heisenberg MPI; Studentenschaft der Universität Basel to Heisenberg (2 Nov 1942) Heisenberg MPI.

580. REM to Heisenberg (21 Oct 1942) Heisenberg MPI.

581. RUL to REM (30 Jul 1942) Uk. H185 III, 7 HUB.

582. NSDAP to Heisenberg (28 Oct 1942) Heisenberg MPI.

583. Werner Heisenberg, "Bericht," (11 Dec 1942) Heisenberg MPI.

584. Deutsche Gesandtschaft to AA (17 Oct 1942) REM 2943, 89 BAP.

585. KWG (Forstmann) to REM (4 Nov 1942) REM 2943, 86 BAP.

586. REM, "Vermerk," (19 Nov 1942) REM 2943, 90 BAP.

587. REM to AA (26 Nov 1942) REM 2943, 93 BAP.

588. Werner Heisenberg, "Bericht," (11 Dec 1942) Heisenberg MPI.

589. Carl Friedrich von Weizsäcker, "Bericht," (around 10 Dec 1942) REM 2943, 97 BAP.

590. Max Planck, "Bericht," (10 Dec 1942) REM 2943, 94 BAP; for further information concerning Planck's foreign lectures, see Heilbron, (1986), 183–91.

591. Deutsche Gesandtschaft (Prag) to AA (6 Jan 1943) REM 2943, 99–100 BAP.

592. For the state of the war see Noakes and Pridham, (1990), 839–45, and Kershaw, (1987), 189–99.

593. Laue to PAW (11 Dec 1942) II–III, 66/7 AdW; SGA II–V, 104 AdW; REM to PAW (17 Apr 1943) II–III, 66/7 AdW.

594. REM to RUB (11 Mar 1943) Uk. H185 II, 8 HUB.

595. Bieberbach to Heisenberg (22 Mar 1943) Uk. H185 II, 8 HUB; Dozentenführer to RUB (17 Mar 1943) Uk. H185 I, 18 HUB.

596. REM to Heisenberg (17 Jul 1943) Heisenberg MPI.

597. Heisenberg to Bieberbach (25 Mar 1943) Uk. H185 I, 20 HUB.

598. Samuel Goudsmit, "Interview with 'F.J.'," (31 Aug 1944) M1108 File 26, National Archives and Records Service (henceforth NAARS).

599. REM to Heisenberg (24 Feb 1943) Heisenberg MPI.

600. Heisenberg to DAAD (9 Mar 1943) Heisenberg MPI.

601. Heisenberg to REM (9 Apr 1943) Heisenberg MPI.

602. Heisenberg to Army Ordnance (13 Jan 43) Heisenberg MPI; Schumann to Heisenberg (18 Jan 43) Heisenberg MPI; Heisenberg to Hahn (22 Jan 43) Heisenberg MPI.

603. Ramsauer to Prandtl (12 May 42, LPG; Carl Ramsauer, "Über Leistung und Organisation der angelsächischen Physik: Mit Ausblicken für die deutsche Physik," (2 Apr 1943) G-241 German Reports, AIP.

604. Abraham Esau, "Einleitung," (5 May 1943) G-214 German Reports, AIP; Abraham Esau, "Herstellung von Leuchtfarben ohne Anwendung von Radium," (5 May 1943) G-213 German Reports, AIP.

605. Otto Hahn, "Künstliche Atomumwandlung und Atomkernspaltung," (5 May 1943) G-216 German Reports, AIP.

606. Klaus Clusius, "Isotopentrennung," (5 May 1943) G-207 German Reports, AIP.

607. Walther Bothe, "Die Forschungsmittel der Kernphysik," (5 May 1943) G-205 German Reports, AIP.

608. Heisenberg, (1989), 570–75.

609. Boettcher to Heisenberg (21 Apr 1943) Heisenberg MPI.

610. Heisenberg to Boettcher (30 Apr 1943) Heisenberg MPI.

611. REM to Heisenberg (15 Jun 1943) Heisenberg MPI.

612. REM to Heisenberg (31 Jul 1943) Heisenberg MPI.

613. Heisenberg to REM (11 Aug 1943) Heisenberg MPI.

614. Kramers to Heisenberg (29 Jul 1943) Heisenberg MPI.

615. Kramers to Heisenberg (5 Jul 1944) Heisenberg MPI.

616. Heisenberg to REM (20 Aug 1943) Heisenberg MPI.

617. Heisenberg to Kramers (20 Aug 1943) Heisenberg MPI.

618. Kramers to Heisenberg (1 Sep 1943) Heisenberg MPI.

619. REM to Heisenberg (6 Sep 1943) Heisenberg MPI; Reichskommissar für die besetzten niederländischen Gebiete to Heisenberg (15 Sep 1943) Heisenberg MPI.

620. Reichskommissar für die besetzten niederländischen Gebiete to Heisenberg (15 Sep 1943) Heisenberg MPI.

621. Cassidy, (1991), 471.

622. Werner Heisenberg, "Bericht," (10 Nov 1943) Heisenberg MPI.

623. Kuiper to Fischer (30 Jun 1945), Gerald Kuiper Papers, Tuscon (henceforth GKT); I would like to thank Ron Doel for showing me this document.

624. Heisenberg to Hiby (1 Nov 1943) Heisenberg MPI.

625. Interview with Stefan Rosental; Rosenfeld to Heisenberg (10 Dec 1943) Heisenberg MPI; Rosenfeld to Heisenberg (14 Apr 1944) Heisenberg MPI.

626. Kramers to Heisenberg (5 Jul 1944) Heisenberg MPI.

627. Heisenberg to Schwarz (14 Feb 1944) Heisenberg MPI.

628. Coblitz to Dennhardt (8 Dec 1943) R 52 IV/152, Bundesarchiv, Koblenz (henceforth BAK).

629. REM to Heisenberg (13 May 1943) Uk. H185 I, 23 HUB.

630. Heisenberg to Coblitz (26 May 1943) Heisenberg MPI.

631. Coblitz to Heisenberg (25 May 1943) Heisenberg MPI; Heisenberg to Coblitz (3 Jun 1943) Heisenberg MPI.

632. For the "Copernicus Prize," see Burleigh, (1988), 279–80.

633. Borger to Heisenberg (7 Jun 1943) Heisenberg MPI.

634. Heisenberg to Borger (11 Jun 1943) Heisenberg MPI.

635. Coblitz to Dennhardt (8 Dec 1943) R 52 IV/152 BAK.

636. Heisenberg to Coblitz (11 Oct 1943) Heisenberg MPI.

637. Heisenberg to Harteck (8 Dec 1943) Heisenberg MPI.

638. Coblitz to Heisenberg (15 Jul 1943) Heisenberg MPI; Coblitz to Heisenberg (29 Sep 1943) Heisenberg MPI; Heisenberg to Coblitz (29

Oct 1943) Heisenberg MPI; Coblitz to Heisenberg (18 Nov 1943) Heisenberg MPI.

639. Cassidy, (1991), 466–67.
640. Burrin, (1994).
641. Spielvogel, (1992), 294.
642. See chapters 8, 9, and 10.
643. REM to Heisenberg (17 Jul 1943) Heisenberg MPI; Heisenberg to Deutsches Wissenschaftliches Institut Bucharest (18 Dec 1943) Heisenberg MPI.
644. Deutsche Akademie to Heisenberg (29 Sep 1943) Heisenberg MPI; Deutsche Akademie to Heisenberg (29 Dec 1943) Heisenberg MPI.
645. REM to RUB (29 Dec 1944) Uk. H185 I, 12 HUB.
646. Laue to Kurator der Uni. Berlin (2 Apr 1944) Uk. H185 I, 11 HUB.
647. "Rapport over Begivenhederne under Besættelsen af Universitetets Institut for teoretisk Fysik fra d.6.December 1943 til d.3.February 1944," 1944 or 1945, NBC; thanks to Finn Aaserud und Gro Næs for translating this document.
648. Von Weizsäcker to Heisenberg (16 Jan 1944) NAARS.
649. "Rapport" NBC.
650. Handwritten note by Heisenberg on back of Euler to Heisenberg (8 Jan 1944) Heisenberg MPI.
651. "Rapport" NBC.
652. Heisenberg to Jensen (1 Feb 1944) Heisenberg MPI.
653. REM to Heisenberg (1 Mar 1944) Heisenberg MPI; Deutsche Forschungsgemeinschaft to Heisenberg (28 Mar 1944) Heisenberg MPI.
654. Werner Heisenberg, "Bericht," (27 Apr 1944) Uk. H185 I, 32 HUB.
655. Wüst to Himmler (15 Oct 1937) Höfler BDC.
656. Otto Höfler BDC.
657. SS-Obersturmführer to Himmler (22 Jun 1937) Höfler BDC.
658. Nagel, (1991).
659. Deichmann, (1992), 199–210.
660. Wüst to Himmler (15 Oct 1937) Höfler BDC.
661. SS-Brigadeführer und Generalmajor der Polizei to dem Reichsführer-SS im Hause (23 Nov 1942) Höfler BDC.
662. Klinger to Heisenberg (28 Jun 1949) Heisenberg MPI.
663. Heisenberg to Klinger (4 Jul 1949) Heisenberg MPI.
664. Heisenberg to Höfler (27 Apr 1944) Heisenberg MPI; Werner Heisenberg, "Bericht," (27 Apr 1944) Uk. H185 I, 32 HUB.
665. Höfler to Heisenberg (12 Jan 1947) Heisenberg MPI.

666. Heisenberg to Klinger (4 Jul 1944) Heisenberg MPI.
667. Werner Heisenberg, "Die aktive und passive Opposition im Dritten Reich," (12 Nov 1947) Heisenberg MPI.
668. REM to Heisenberg (9 Aug 1944) Uk. H185 I, 35 HUB.
669. Powers, (1993), 402.
670. Goudsmit, (1983), 114.
671. REM to Heisenberg (27 Mar 1945) Uk. H185 I, 36 HUB.
672. For the ambivalence towards Heisenberg and von Weizsäcker, see Walker, (1989), 204–21.
673. See chapter 10.
674. Kleinert, (1979); Hentschel, (1990); Hentschel, *Studies*, (1992); Hentschel, *Turm*, (1992).
675. Forman, (1973).
676. Forman, (1974).
677. Forman, (1973); Forman, (1974); Forman, "Helmholtz."
678. Noakes and Pridham, (1990), 697–99.
679. Noakes and Pridham, (1990), 220–32.
680. The classic accounts are Ludwig and Beyerchen, (1977), 9–50; also see Mehrtens, (1989) and Geuter (1992).
681. See Kershaw, (1991), 37–86 and the primary documents collected and edited by Noakes and Pridham, (1990), 123–91.
682. Noakes and Pridham, (1990), 123–87, esp. 144–54.
683. Noakes and Pridham, (1990), 146, 148–50.
684. Noakes and Pridham, (1990), 170–1.
685. Noakes and Pridham, (1990), 167–87.
686. Renneberg and Walker, in Renneberg and Walker, (1993), 9–11.
687. Richter, (1973); Beyerchen, (1977); Kleinert, (1978); Richter, (1978/1979); Kleinert, (1980), 35–38; Richter, (1980); Kleinert in Olff-Nathan, (1993).
688. Heisenberg, (1973), 206–7.
689. Neufeld in Renneberg and Walker, (1993); Neufeld, (1994).
690. Neufeld in Renneberg and Walker, (1993).
691. Heisenberg, (1973), 214; Heisenberg, (1946), 329.
692. Heisenberg, (1946), 329.
693. Noakes and Pridham, (1990), 839–45.
694. Walker, (1989), 209.
695. Vollnhals, (1991).
696. Nietzhammer, (1982).
697. Vollnhals, (1991), 57.

698. See Wise, (1993) and Beyler (1994).

699. Walker, (1989), 198–99.

700. See chapters 2 and 3.

701. Gimbel, (1986); Gimbel, (1990).

702. Albrecht, Heinemann-Grüder, and Wellmann, (1992).

703. See Walker, (1989), and chapter 8.

704. Unfortunately, neither the recordings themselves nor the German originals of the recorded conversations are available; the transcripts have been published, including the two sections available in the original German, as Charles Frank (ed.), *Operation Epsilon*, (1993).

705. Groves, (1983), 333–40.

706. Goudsmit, (1983), 134–9.

707. The information on membership in National Socialist organizations is available in the personnel files of these scientists at the BDC.

708. Also see chapter 7.

709. *Operation Epsilon*, (1993), 39, 42, 54.

710. *Operation Epsilon*, (1993), 54, 215.

711. *Operation Epsilon*, (1993), 51–3.

712. *Operation Epsilon*, (1993), 52–3, 64–5, 189–90.

713. *Operation Epsilon*, (1993), 50, 168, 171.

714. *Operation Epsilon*, (1993), 50, 55.

715. Walker, (1989), 204–221.

716. Walker, (1989), 21, 56–58.

717. *Operation Epsilon*, (1993), 72, 109, 117–21.

718. Walker, (1989), 48–9.

719. Walker, (1989), 46–51, 165–78.

720. *Operation Epsilon*, (1993), 70.

721. *Operation Epsilon*, (1993), 70–1.

722. Walker, (1989), 23.

723. *Operation Epsilon*, (1993), 71–2.

724. *Operation Epsilon*, (1993), 72–4.

725. *Operation Epsilon*, (1993), 74.

726. Walker, (1989), 23–24.

727. *Operation Epsilon*, (1993), 74.

728. *Operation Epsilon*, (1993), 75.

729. *Operation Epsilon*, (1993), 76.

730. *Operation Epsilon*, (1993), 75–7, 83.

731. *Operation Epsilon*, (1993), 83.

732. Walker, (1989), 96–9.

733. *Operation Epsilon*, (1993), 83–4.
734. *Operation Epsilon*, (1993), 88.
735. Walker, (1989), 21.
736. *Operation Epsilon*, (1993), 116–17.
737. *Operation Epsilon*, (1993), 117–18.
738. *Operation Epsilon*, (1993), 118–40; the original German version of this lecture is given on pages 147–64.
739. *Operation Epsilon*, (1993), 143–4.
740. *Operation Epsilon*, (1993), 175.
741. Walker, (1989), 26–7, 168, 207, 209.
742. Smyth, (1945).
743. Walker, (1989), 25–27.
744. *Operation Epsilon*, (1993), 116, 119–20.
745. *Operation Epsilon*, (1993), 117–18.
746. R.V. Jones, Introduction, in Goudsmit, (1983), xv-xvi; also see Frank's introduction to *Operation Epsilon*, (1993), 1–13, especially 5–7.
747. Walker, (1989), 49–51.
748. Goudsmit, (1983), xvi.
749. *Operation Epsilon*, (1993), 72.
750. *Operation Epsilon*, (1993), 70, 82.
751. *Operation Epsilon*, (1993), 79–80.
752. *Operation Epsilon*, (1993), 80.
753. *Operation Epsilon*, (1993), 82.
754. *Operation Epsilon*, (1993), 87.
755. *Operation Epsilon*, (1993), 72.
756. *Operation Epsilon*, (1993), 79.
757. *Operation Epsilon*, (1993), 73, 78, 85.
758. *Operation Epsilon*, (1993), 85–6, 90.
759. *Operation Epsilon*, (1993), 76–7.
760. *Operation Epsilon*, (1993), 78.
761. *Operation Epsilon*, (1993), 80, 85–6.
762. *Operation Epsilon*, (1993), 78.
763. *Operation Epsilon*, (1993), 76–7.
764. *Operation Epsilon*, (1993), 90.
765. *Operation Epsilon*, (1993), 78.
766. *Operation Epsilon*, (1993), 83.
767. *Operation Epsilon*, (1993), 67.
768. *Operation Epsilon*, (1993), 82–3.
769. Walker, (1989), 46–51, 205–11.

770. *Operation Epsilon*, (1993), 175.
771. *Operation Epsilon*, (1993), 92.
772. Jungk, (1958), 105.
773. *Operation Epsilon*, (1993), 79–80.
774. *Operation Epsilon*, (1993), 92.
775. *Operation Epsilon*, (1993), 171.
776. *Operation Epsilon*, (1993), 27.
777. *Operation Epsilon*, (1993), 143.
778. *Operation Epsilon*, (1993), 89.
779. *Operation Epsilon*, (1993), 145.
780. See chapter 7.
781. *Operation Epsilon*, (1993), 87.
782. *Operation Epsilon*, (1993), 87–8.
783. Walker, (1989), 192–221.
784. *Operation Epsilon*, (1993), 143.
785. *Operation Epsilon*, (1993), 88–9.
786. *Operation Epsilon*, (1993), 92–3.
787. *Operation Epsilon*, (1993), 92.
788. *Operation Epsilon*, (1993), 211.
789. Walker, (1989), 177.
790. *Operation Epsilon*, (1993), 88.
791. *Operation Epsilon*, (1993), 143.
792. *Operation Epsilon*, (1993), 110, 143.
793. See for example Eckert, (1990) and Eckert and Osietzki, (1989).
794. *Operation Epsilon*, (1993), 108–9.
795. *Operation Epsilon*, (1993), 109.
796. *Operation Epsilon*, (1993), 168.
797. *Operation Epsilon*, (1993), 230.
798. *Operation Epsilon*, (1993), 236.
799. *Operation Epsilon*, (1993), 241.
800. Walker, (1989), 204–21.
801. *Operation Epsilon*, (1993), 179–80.
802. *Operation Epsilon*, (1993), 221.
803. Walker, (1989), 179–204.
804. *Operation Epsilon*, (1993), 266.
805. *Operation Epsilon*, (1993), 275.
806. *Operation Epsilon*, (1993), 190.
807. *Operation Epsilon*, (1993), 77.
808. Walker, (1989), 192–221.

809. Walker, (1989), 153–60.
810. Goudsmit, (1983).
811. Jungk, (1956).
812. Cioc, (1988), 43–6, 75–91; an English version of the manifesto was reprinted as "Declaration ...," (1957).
813. Von Weizsäcker, (1957).
814. See chapters 6 and 7.
815. Ladenburg to Goudsmit (23 Oct 1946) Goudsmit, AIP.
816. Van der Waerden to Heisenberg (18 Mar 1948) Heisenberg MPI.
817. Heisenberg to van der Waerden (28 Apr 1948) AIP.
818. Jungk, (1956), 108–11, esp. 110.
819. Jungk to Heisenberg (10 Feb 1955) Heisenberg MPI; Heisenberg to Jungk (14 Feb 1955) Heisenberg MPI.
820. Heisenberg to Jungk (17 Nov 1956) Heisenberg MPI.
821. Jungk, (1958) 101; Heisenberg to Jungk (18 Jan 1957) Heisenberg MPI.
822. Jungk, (1958), 102–4.
823. Jungk, (1958), 81, 91, 102–4; Heisenberg to Jungk (18 Jan 1957) Heisenberg MPI; Heisenberg to Jungk (17 Nov 1956) Heisenberg MPI.
824. After this passage had been written, Jungk corroborated this interpretation; Jungk to the author (30 Apr 1989).
825. Von Weizsäcker to the editor (14 Oct 1955), in Seelig, 130–33, here 130–31.
826. Von Weizsäcker, (1958), 180–81.
827. Bohr, (1967), 193; see chapters 6 and 7.
828. Kramish, (1985), 120, 203.
829. Feinberg, (1989); I would like to thank Carl Friedrich von Weizsäcker for sending me a German translation of Feinberg's article.
830. Citations are from Heisenberg, (1973), 211–14.
831. Citations are from Heisenberg, (1983), 93.
832. Heisenberg, (1983), 96–103.
833. Jungk, (1993), 298–300.
834. Von Weizsäcker, *Welt*, (1991); von Weizsäcker, *Spiegel*, (1991); also see von Weizsäcker, (1988), 37–83.
835. See von Weizsäcker, (1993), 331–60, here 351.
836. Von Weizsäcker, *Welt*, (1991), 4 .
837. Von Weizsäcker, *Zeit*, (1991), 18.
838. Walker, (1989), 47–51.
839. *Newsweek* (7 Jul 1958), 62; Diamond to Goudsmit (5 Jul 1958) AIP.
840. See chapter 9.

841. Groves, (1983), 333–40; see chapter 9.
842. Goudsmit, (1983), 134–9.
843. The latest in this line are Kramish, (1985) and Powers, (1993).
844. David Irving's book, although flawed in other ways, is a notable exception; see Irving, (1983).
845. Powers, (1993), 479; the page references here are taken from the British edition published by Jonathan Cape.
846. Powers, (1993), 507.
847. See chapters 6 and 7.
848. See chapter 9.
849. Powers, (1993), 481.
850. See Hermann and Schumacher, (1986), 11–22.
851. Kramish, (1985), 118–9.
852. Kershaw, (1993), 197–217.
853. Gillispie, (1959).
854. Graham, (1993).

Bibliography

Helmuth Albrecht (ed.), *Naturwissenschaft und Technik in der Geschichte. 25 Jahre Lehrstuhl für Geschichte der Naturwissenschaft und Technik am Historischen Institut der Universität Stuttgart* (Stuttgart, GNT Verlag, 1993)

Helmuth Albrecht, "'Max Planck: Mein Besuch bei Adolf Hitler'— Anmerkungen zum Wert einer historischen Quelle," in Albrecht (1993), 41–63

Helmuth Albrecht and Armin Hermann, "Die Kaiser-Wlhelm-Gesellschaft im Dritten Reich," in Vierhaus and Brocke (1990), 356–406

Ulrich Albrecht, Andreas Heinemann-Grüder, and Arend Wellmann, *Die Spezialisten. Deutsche Naturwissenschaftler und Techniker in der Sowjetunion nach 1945* (Berlin, Dietz, 1992)

Alan Beyerchen, *Scientists under Hitler: Politics and the Physics Community in the Third Reich* (New Haven, Yale University Press, 1977)

Alan Beyerchen, "What we now know about Nazism and science," *Social Research*, 59 (1992), 615–41

Richard Beyler, "From Positivism to Organicism: Pascual Jordan's Interpretations of Modern Physics in Cultural Context," (Harvard University Ph.D., 1984).

Aage Bohr, "The War Years and the Prospects Raised by the Atomic Weapons," in Stefan Rozental (1967)

Michael Burleigh, *Germany Turns Eastward: A Study of Ostforschung in the Third Reich* (Cambridge, Cambridge University Press, 1988)

Philippe Burrin, *Hitler and the Jews: The Genesis of the Holocaust* (London, Edward Arnold, 1994)

David Cassidy, *Uncertainty: The Life and Science of Werner Heisenberg* (New York, Freeman, 1991)

Marc Cioc, *Pax Atomica: The Nuclear Defense Debate in West Germany during the Adenauer Era* (New York, Columbia University Press, 1988)

J. G. Crowther, *Science in Liberated Europe* (London, Pilot 1949)

"Declaration of the German Nuclear Physicists," *Bulletin of the Atomic Scientists*, 8, No. 6 (1957), 228

Ute Deichmann, *Biologen unter Hitler. Vertreibung, Karrieren, Forschung* (Frankfurt a. M., Campus Verlag, 1992)

Michael Eckert, "Primacy doomed to failure: Heisenberg's role as scientific advisor for nuclear policy in the FRG," *Historical Studies in the Physical and Biological Sciences*, 21 (1990), 29–58

Michael Eckert, *Die Atomphysiker. Eine Geschichte der theoretischen Physik am Beispiel der Sommerfeldschule* (Braunschweig, Vieweg, 1992)

Michael Eckert and Maria Osietzki, *Wissenschaft für Macht und Markt. Kernforschung und Mikroelektronik in der Bundesrepublik Deutschland* (Munich, Beck Verlag, 1989)

Eugene Feinberg, "Werner Heisenberg: The Tragedy of a Scientist" [Russian] *Znamja*, Nr. 3 (1989) 124–43

Klaus Fischer, "Die Emigration von Wissenschaftlern nach 1933. Möglichkeiten und Grenzen einer Bilanzierung," *Vierteljahrshefte für Zeitgeschichte*, 39 (1991), 535–49

Paul Forman, "Scientific Internationalism and the Weimar Physicists: The Ideology and its Manipulation in Germany after World War I," *Isis*, 64 (1973), 151–80

Paul Forman, "The Financial Support and Political Alignment of Physicists in Weimar Germany," *Minerva*, 12 (1974), 39–66

Paul Forman, "The Helmholtz Gesellschaft" (manuscript).

Paul Forman, "The Naturforscherversammlung in Nauheim, September 1920: An Introduction to Scientific Life in the Weimar Republic" (manuscript)

Charles Frank, *Operation Epsilon: The Farm Hall Transcripts* (Berkeley, University of California Press, 1993)

Ulfried Geuter, *The Professionalization of Psychology in Nazi Germany* (Cambridge, Cambridge University Press, 1992)

Charles C. Gillispie, "The *Encyclopédie* and the Jacobin Philosophy of Science: A Study in Ideas and Consequences," in Marshall Clagett (ed.), *Critical Problems in the History of Science* (Madison, University of Wisconsin Press, 1959), 255–289

John Gimbel, "U.S. Policy and German Scientists: the Early Cold War," *Political Scien.. Quarterly*, 101 (1986), 433–51

John Gimbel, *Science, Technology, and Reparations. Exploitation and Plunder in Postwar Germany* (Palo Alto, Stanford University Press, 1990)

Ludwig Glaser, "Ueber Versuche zur Bestätigung der Relativitätstheorie an der Beobachtung," *Glasers Annalen für Gewerbe und Bauwesen* (15 August 1920), No. 1036, 29–33

Ludwig Glaser, "Jüdischer Geist in der Physik," *Zeitschrift für die gesamte Naturwissenschaft*, 5 (1939), 162–75

Ludwig Glaser, "Die Sommerfeldsche Feinstruktur als prinzipielle Frage der Physik," *Zeitschrift für die gesamte Naturwissenschaft*, 5 (1939), 289–331

Joseph Goebbels, *The Goebbels Diaries 1942–1943* (New York, Doubleday, 1948)

Samuel Goudsmit, *Alsos* 2nd ed. (New York, Tomash, 1983)

Loren Graham, *Science in Russia and the Soviet Union: A Short History* (Cambridge, Cambridge University Press, 1993)

Conrad Grau, Wolfgang Schlicker, and Liane Zeil, *Die Berliner Akademie der Wissenschaften in der Zeit des Imperialismus. Teil III: die Jahre der faschistischen Diktatur 1933 bis 1945* (Berlin, Akademie-Verlag, 1979)

Leslie Groves, *Now It Can Be Told* 2nd Ed. (New York, Da Capo, 1983)

Joseph Haberer, *Politics and the Community of Science* (New York, Van Nostrand, 1969)

John L. Heilbron, *The Dilemmas of an Upright Man: Max Planck as Spokesman for German Science* (Berkeley, University of California Press, 1986)

Rudolf Heinrich and Hans-Reinhard Bachmann (eds.), *Walther Gerlach. Physiker–Lehrer–Organisator* (Munich, Deutsches Museum, 1989)

Elisabeth Heisenberg, *Das politische Leben eines Unpolitischen. Erinnerungen an Werner Heisenberg* (Munich, 1983)

Werner Heisenberg, "Die Bewertung der 'modernen theoretischen Physik,'" *Zeitschrift für die gesamte Naturwissenschaft*, 9 (1943), 201–12

Werner Heisenberg, "über die Arbeiten zur technischen Ausnutzung der Atomkernenergie in Deutschland," *Die Naturwissenschaften*, 33 (1946), 325–29

Werner Heisenberg, *Der Teil und das Ganze. Gespräche im Umkries der Atomphysik* 2nd ed. (Munich, Deutscher Taschenbuch Verlag, 1973)

Werner Heisenberg, *Gesammelte Werke/Collected Works, Volume A II* (Berlin, Springer, 1989)

Klaus Hentschel, *Interpretationen und Fehlinterpretationen der speziellen und allgemeinen Relativitätstheorie durch Zeitgenossen Albert Einsteins* (Basel, Birkhäuser, 1990)

Klaus Hentschel, "Grebe/Bachems photomechanische Analyse der Linienprofile und die Gravitations-Rotversheibung: 1919–1922," *Annals of Science*, 49 (1992), 21-46

Klaus Hentschel, "Einstein's attitude towards experiments—testing relativity theory, 1907–1927," *Studies in History and Philosophy of Science*, 23 (1992), 593–624

Klaus Hentschel, *Der Einstein-Turm* (Heidelberg, Spektrum, 1992)

Klaus Hentschel and Monika Renneberg, "Ausschaltung oder 'Verteidigung' der allgemeinen Relativitätstheorie—Interpretationen einer Kosmologen-Karriere im Nazisozialismus," in Meinel and Voswinckel, (1994), 201-207

Armin Hermann, *The New Physics, The Route into the Atomic Age* (Bonn, Inter Nationales, 1979)

Armin Hermann and Rolf Schumacher (eds.), *Das Ende des Atomzeitalters?* (Munich, Moos, 1986)

Dieter Hoffmann, "Johannes Stark—eine Persönichkeit im Spannungsfeld von wissenschaftlicher Forschung und faschistischer Ideologie,"

Philosophie und Naturwissenschaften in Vergangenheit und Gegenwart, 22 (1982), 90–101

Dieter Hoffmann, "Zur Teilnahme deutscher Physiker an der kopenhagener Physikerkonferenzen nach 1933 sowie am 2 Kongress für Einheit der Wissenschaft, Kopenhagen 1936," *NTM,* 25 (1988), 49–55

Dieter Hoffmann, "Die Physikdenkschriften von 1934/36 und zur Situation der Physik im faschistischen Deutschland," *XVIII Internationaler Kongress für Geschichte der Wissenschaften* (Berlin, Akademie der Wissenschaften der DDR, 1989), 185–211

Dieter Hoffmann, "Walter Nernst und die Physikalisch-Technische Reichsanstalt," in *PTB-Mitteilungen,* 100 (1990), 40–45

Dieter Hoffmann, "Nationalsozialistische Gleichschaltung und Tendenzen militärtechnischer Forschungsorientierung an der Physikalisch-Technischen Reichsanstalt im Dritten Reich," in Albrecht (1993), 121–31

Dieter Hoffmann (ed.), *Operation Epsilon. Die Farm-Hall-Protokolle oder die Angst der Alliierten vor der deutschen Atombombe* (Berlin, Rowohlt Berlin, 1993)

David Irving, *The German Atomic Bomb: The History of Nuclear Research in Nazi Germany* 2nd ed. (New York, Da Capo, 1983)

Robert Jungk, *Heller als tausend Sonnen* (Stuttgart, Scherz, 1956)

Robert Jungk, *Brighter Than a Thousand Suns* (New York, Harcourt, Brace, and Company, 1958)

Robert Jungk, *Trotzdem. Mein Leben für die Zukunft* (Munich, Hanser, 1993)

Ian Kershaw, *The "Hitler Myth": Image and Reality in the Third Reich* (Oxford, Oxford University Press, 1987)

Ian Kershaw, *Hitler* (London, Longman, 1991)

Ian Kershaw, *The Nazi Dictatorship: Problems and Perspectives of Interpretation* 3rd ed. (London, 1993)

Andreas Kleinert, "Von der science allemande zur deutschen Physik," *Francia,* 6 (1978), 509–25

Andreas Kleinert, "Nationalistische und antisemitische Ressentiments von Wissenschaftlern gegen Einstein," in H. Nelkowski et. al. (eds.), *Einstein Symposium Berlin.* (Berlin, Springer Verlag, 1979), 501–16

Andreas Kleinert, "Lenard, Stark und die Kaiser-Wilhelm-Gesellschaft. Auszüge aus der Korrespondenz der beiden Physiker zwischen 1933 und 1936," *Physikalische Blätter*, 36, No. 2, (1980), 35–43

Andreas Kleinert, "Das Spruchkammerverfahren gegen Johannes Stark," *Sudhoffs Archiv*, 67 (1983), 13–24

Andreas Kleinert, "La correspondance entre Philipp Lenard et Johannes Stark," in Olff-Nathan (1993), 149–66

Andreas Kleinert, "Paul Weyland, der Berliner Einstein-Töter," in Albrecht (1993), 198–232

Arnold Kramish, *The Griffin* (Boston, Houghton Mifflin, 1985)

Max von Laue, *Die Naturwissenschaften*, 11 (1923), 29

Max von Laue, "Bemerkung zu der vorstehenden Veröffentlichung von J. Stark," *Physikalische Blätter*, 3 (1947), 272–73

Max von Laue, "Ansprache bei Eröffnung der Physikertagung in Würzburg am 18. September 1933," in Laue, *Gesammelte Schriften und Vorträge*, *Volume III* (Braunschweig, Vieweg, 1961), 61–62

Philipp Lenard, *Deutsche Physik*

Karl-Heinz Ludwig, *Technik und Ingenieure im Dritten Reich* 2nd ed. (Düsseldorf, Droste, 1979)

Kristie Macrakis, *Surviving the Swastika: Scientific research in Nazi Germany* (New York, Oxford University Press, 1993)

Herbert Mehrtens, "Angewandte Mathematik und Anwendung der Mathematik im nationalsozialistischen Deutschland," *Geschichte und Gesellschaft*, 12 (1986), 317–47

Herbert Mehrtens, "The 'Gleichschaltung' of Mathematical Societies in Nazi Germany," *The Mathematical Intelligencer*, 11 (1989), 48–60

Herbert Mehrtens, "Irresponsible Purity: the Political and Moral Structure of Mathematical Sciences in the National Socialist State," in Renneberg and Walker (1993), 324–38 & 411–13

Herbert Mehrtens, "Kollaborationsverhältnisse: Natur- und Technikwissenschaften im NS-Staat und ihre Histoire," in Meinel and Voswinckel, (1994), 13-32

Herbert Mehrtens, "The Social System of Mathematics and National Socialism," in Renneberg and Walker (1993), 291–311 & 406–8

Bibliography   313

Herbert Mehrtens and Steffen Richter (eds.), *Naturwissenschaft, Technik und Ideologie: Beiträge zur Wissenschaftsgeschichte des Dritten Reiches* (Frankfurt a. M., Suhrkamp, 1980)

Christoph Meinel and Peter Voswinckel (eds.), *Medizin, Naturwissenschaften, Technik un Nazionalsozialismus. Kontinuitäten und Diskontinuitäten* (Stuttgart, GNT Verlag, 1994)

Willi Menzel, "Deutsche Physik und jüdische Physik," *Völkischer Beobachter, Norddeutsche Ausgabe*, (29 Jan 1936)

Wilhelm Müller, "Jüdischer Geist in der Physik," *Zeitschrift für die gesamte Naturwissenschaft*, 5 (1939), 162–75

Nevill Mott and Rudolf Peierls, "Werner Heisenberg: 5 Dec 1901–1 February 1976," *Biographical Memoirs of Fellows of the Royal Society*, 23 (1977), 236.

Brigitte Nagel, *Die Welteislehre. Ihre Geschichte und ihre Rolle im "Dritten Reich"* (Stuttgart, GNT-Verlag, 1991)

Michael Neufeld, "The Guided Missile and the Third Reich: Peenemünde and the Forging of a Technological Revolution," in Renneberg and Walker, (1993), 51–71 & 352–56

Michael Neufeld, *The Rocket and the Reich. Peenemüde and the Coming of the Ballistic Missile Era* (New York, Free Press, 1995)

Lutz Nietzhammer, *Die Mitläuferfabrik. Die Entnazifizierung am Beispiel Bayerns* 2nd ed. (Bonn, Dietz, 1982)

Jeremy Noakes and Geoffrey Pridham (eds.), *Nazism: A History in Documents and Eyewitness Accounts, 1919–1945. Volumes 1 & 2* (New York, Schocken, 1990)

Josiane Olff-Nathan (ed.), *La Science sous le Troisième Reich* (Paris, Le Seuil, 1993)

Josiane Olff-Nathan, "Nazisme et science: un mariage de Raison?" in Éric Heilmann (ed.), *Science ou justice? Les savants, l'ordre et la roi* (Paris, Éditions Autrement, 1994), 132–48

Josiane Olff-Nathan, "Science de fous ou folie de la science," *Cliniques Méditerranéenes*, 41/42 (1994), 103–13

Thomas Powers, *Heisenberg's War: The Secret History of the German Bomb* (New York, Knopf, 1993)

Monika Renneberg, "Zur mathematischen-Naturwissenschaftlichen Fakultät der Hamburger Universität im 'Dritten Reich'," in Eckart

Krause, Ludwig Huber, and Holger Fischer (eds.), *Hochschulalltag im "Dritten Reich." Die Hamburger Universität 1933–1945* (Berlin, Reimer, 1991), 1051-71

Monika Renneberg, "Die Physik und die physikalischen Institute an der Hamberger Universität im 'Dritten Reich'," in Krause, Huber, and Fischer, 1097-118

Monika Renneberg and Mark Walker (eds.), *Science, Technology, and National Socialism* (Cambridge, Cambridge University Press, 1993)

Monika Renneberg and Mark Walker, "Scientists, Engineers, and National Socialism," in Renneberg and Walker, (1993), 1–29 & 339–46

Steffen Richter, "Die Kämpfe innerhalb der Physik in Deutschland nach dem Ersten Weltkrieg," *Sudhoffs Archiv*, 57 (1973), 195–207

Steffen Richter, "Physiker im Dritten Reich," *Jahrbuch der TH Darmstadt* (1978/1979), 103–13

Steffen Richter, "Die 'Deutsche Physik'" in Mehrtens and Richter (1980), 116–41

Stefan Rozental (ed.), *Niels Bohr* (Amsterdam, Nortj-Holland, 1967)

H. Rügemer, "Die 'Nature' eine Greuelzeitschrift," *Zeitschrift für die gesamte Naturwissenschaft*, 3 (1937), 475–79

Brigitte Schroeder-Gudehus, "The Argument for Self-Government and Public Support of Science in Weimar Germany," *Minerva*, 10 (1972), 537–70

Brigitte Schroeder-Gudehus, "Challenge to Transnational Loyalties: International Scientific Organizations after World War I," *Science Studies*, 3 (1973), 93–118

Brigitte Schroeder-Gudehus, *Les Scientifiques et la Paix. La Communauté Scientifique Internationale au Cours des Années 20* (Montreal, Les Presses de l'Université de Montréal, 1978)

Carl Seelig (ed.), *Helle Zeit—Dunkle Zeit* (Zurich, Europa Verlag, 1956)

Reinhard Siegmund-Schultze, "Theodor Vahlen—zum Schuldanteil eines deutschen Mathematikers am faschistischen Missbrauch der Wissenschaft," *NTM*, 21 (1984), 117–32

Henry Smyth, *Atomic Energy for Military Purposes* (Princeton, Princeton University Press, 1945)

Jackson Spielvogel, *Hitler and Nazi Germany: A History* 2nd ed. (Englewood Cliffs, Prentice Hall, 1992)

Johannes Stark, "International Status and Obligations of Science," *Nature* (24 Feb 1934), 290

Johannes Stark, "The Attitude of the German Government towards Science," *Nature* (21 Apr 1934), 614

Johannes Stark, "Stellungsnahme von Prof. Dr. J. Stark," *Völkischer Beobachter* (28 Feb 1936), 6

Johannes Stark, "Die 'Wissenschaft' versagte politisch," *Das Schwarze Korps* (15 Jul 1937), 6

Johannes Stark, "The Pragmatic and Dogmatic Spirit in Physics," *Nature*, 141 (30 Apr 1938), 770–72

Johannes Stark, "Zu den Kämpfen in der Physik während der Hitler-Zeit," *Physikalsiche Blätter*, 3 (1947), 271–2

Johannes Stark, *Erinnerungen eines deutschen Naturforschers* (Mannheim, Bionomica Verlag, 1987)

Helmuth Trischler, "Self-Mobilization or Resistenz? The Behavioral Pattern of Aeronautic Scientists before, during, and after National Socialism," in Renneberg and Walker (1993), 72–87 & 356–58

Rudolf Vierhaus and Bernhard vom Brocke (eds.), *Forschung im Spannungsfeld von Politik und Gesellschaft—Geschichte und Struktur der Kaiser-Wilhelm/Max-Planck-Gesellschaft* (Stuttgart, DVA, 1990)

Clemens Vollnhals, *Entnazifizierung. Politische Säuberung und Rehabilitierung in den vier Besatzungszonen 1945–1949* (Munich, DTV, 1991)

Mark Walker, *German National Socialism and the Quest for Nuclear Power, 1939–1949* (Cambridge, Cambridge University Press, 1989)

Max Weinreich, *Hitler's Professors: The Part of Scholarship in Germany's Crimes against the Jewish People* (New York, YIVO, 1946)

"'Weisse Juden' in der Wissenschaft," *Das Schwarze Korps* (15 Jul 1937), 6

Carl Friedrich von Weizsäcker, "Should Germany Have Atomic Arms?" *Bulletin of the Atomic Scientists*, 8, No. 7 (1957), 283–86

Carl Friedrich von Weizsäcker, "Do We Want to Save Ourselves?" *Bulletin of the Atomic Scientists*, 9, No. 5 (1958), 180–84

Carl Friedrich von Weizsäcker, *Bewusstseinswandel* (Munich, Hanser, 1988)

Carl Friedrich von Weizsäcker, "Interview," *Der Spiegel*, (22 April 1991), 227–38

Carl Friedrich von Weizsäcker, "Interview," *Die Welt*, (21 May 1991), 7 & (23 May 1991), 4

Carl Friedrich von Weizsäcker, "Letter to the Editor," *Die Zeit*, (24 May 1991), 18

Carl Friedrich von Weizsäcker, "Interview," in Hoffmann, *Operation*, (1993), 331–60

M. Norton Wise, "Pascual Jordan: Quantum Mechanics, Psychology, National Socialism," in Renneberg and Walker (1993), 224–54, & 391–96

Index

317